The Middle Ages on Television

The Middle Ages on Television

Critical Essays

Edited by
Meriem Pagès *and*
Karolyn Kinane

McFarland & Company, Inc., Publishers
Jefferson, North Carolina

Library of Congress Cataloguing-in-Publication Data

The Middle Ages on television : critical essays / edited by Meriem Pagès and Karolyn Kinane.
p. cm.
Includes bibliographical references and index.

ISBN 978-0-7864-7941-2 (softcover : acid free paper) ∞
ISBN 978-1-4766-2009-1 (ebook)

1. Middle Ages on television. 2. Medievalism on television. I. Pagès, Meriem, editor. II. Kinane, Karolyn, editor.

PN1995.9.M52M53 2015
791.45'658207—dc23 2015007676

British Library cataloguing data are available

Printed in the United States of America

McFarland & Company, Inc., Publishers
Box 611, Jefferson, North Carolina 28640
www.mcfarlandpub.com

Table of Contents

film, whose use of the medieval has been examined repeatedly by scholars—examples of such recent scholarship on "medieval" films include *A Knight at the Movies* (2003), *The Reel Middle Ages* (2006), and *The Medieval Hero on Screen* (2004)—television medievalisms have garnered scarcely any critical attention. On the rare occasion when they have found themselves subject to critical analysis, television medievalisms have been treated as essentially similar to cinematic ones. The excellent volume *The Medieval Hero on Screen* (2004), for instance, explores various popular medievalisms in cinema and television without distinguishing between the two media. The goal of the present collection is to fill this gap: we hope to draw our readers' attention to the existence of television medievalism, a subdivision of popular medievalism different from other kinds of contemporary reimaginings of the Middle Ages—including but not restricted to other visual media such as film and video games.[2] Secondly, we seek, both in this introduction and in the essays that comprise our collection, to begin to sketch the outline of popular television medievalism and identify some of its unique characteristics.

Before launching into a discussion of television medievalism, we must first examine the unlikely alliance between medievalism and the commercial medium of television. The question of the relationship between medievalism and television is more loaded than might, at first, seem to be the case. Loosely defined by Ute Berns and Andrew James Johnston as "the investigation into different ways in which the Middle Ages have been perceived and constructed by later periods,"[3] medievalism has a long and healthy history going back at least to the 18th century. Throughout most, if not all, of that history, the medieval was reimagined almost endlessly to serve various contemporary ends, ones with ideological overtones. "In the 19th century," John M. Ganim argues, "medievalism was constructed as a fierce reproach as well as a utopian escape from the present, and that reproach was framed in explicitly political terms."[4] Ganim presents William Morris as an example of a 19th-century scholar and artist whose use of the medieval encompassed both the realm of fantasy and that of political and social activism.[5] In the middle of the 20th century, the opposite became true, with the Middle Ages reinforcing conservative values after the devastation of World War II and during the Cold War.[6] Today, our perception of the medieval has come to acquire new meaning and to serve yet another function: in a 21st-century society that no longer believes in the effectiveness and fundamental integrity of the nation-state or in the Renaissance myth of continuous growth and progress, the Middle Ages is starting to displace the Renaissance as a point of origin for the modern period.[7] In Steve Guthrie's words, "Popular fiction images of the Middle Ages focus on personal trial as a training ground in times of crises, when the central authority of the nation-state is absent or failed and the world has gone local, without the Renaissance beliefs in statecraft and progress to hold it together."[8] The medieval world is beginning to look, sound, and feel more comfortable; rather than a time and place full of cruelty, plague, and bad teeth, the Middle Ages now bears an uncanny resemblance to our own day.

If medievalism is influenced at least in part by ideological considerations, the same cannot so easily be said of television programming. Television usually operates according to strict capitalistic principles: its goal lies in bombarding its viewers with images that will entice them to purchase the products featured on the channel's publicity spots, airtime itself obtained as the result of a financial exchange between the channel and its client. Looking at television from this purely consumerist perspective, one might argue that television shows, "medieval" and otherwise, exist simply to ensure that audiences remain entertained enough not to change channels before the first commercial break. The more captive the viewers, the more likely they are to watch advertisements while waiting for their favorite program to resume and the greater the chance that they will buy the goods publicized.

At the same time, most channels develop a set of characteristics, a personality of sorts, that allows them to attract specific audiences for whom advertisements—and, of course, programs—are then tailored. FOX, HBO, and PBS, for example, are very distinct from one another; each of these channels has become known for a particular type of programming with clear ideological implications. Although less in evidence than consumerism, ideology, then, is not completely absent from the production of medieval-themed (and other) shows. While a channel's capitalist needs often dictate its ideology, especially in the case of pay-television channels such as HBO, the latter greatly influences ad selection and placement on other occasions—for instance, ideology plays an important role in determining programming on FOX. If ideological concerns take an active part in shaping the creative decisions that result in a given television program, however, these are closely intertwined with consumerism on most, if not all, such visual products.

In this, television stands apart from its older and more "serious" sibling, cinema. Although both commercial film and television share the need to entertain their audiences visually, their relationship with the consumerist, capitalist culture that produces them differs considerably one from the other. Even the most unabashedly commercial film does not present the kind of marketing opportunities available on television. Rather, much thought must go into the marketability of such concoctions, with entire teams dedicated to creating casts of characters that will translate into attractive supermarket shelf toys—and the much cheaper Happy Meal versions that will entice children to make their parents buy them the real thing. Angela Jane Weisl has shown, for example, the ways in which the *Star Wars* franchise replicates itself endlessly simply to generate more toys and greater marketability.[9] But even *Star Wars* cannot sell generic products effortlessly. This remains the prerogative of commercial television with its format of hour-long programs within which are embedded four publicity breaks; only with such a format in place can a show set against the background of 14th-century England sell soap.

What about networks such as HBO, Starz, and Showtime, channels that pride themselves on precisely their lack of commercials? Pay-television presents yet another approach to consumerism, one that appears closer to that embraced by commercial film than free-to-air television. Like FTA television, HBO and its ilk aim to sell: cable networks have to render themselves appealing to customers who believe they are purchasing a "special" commodity marked by high-quality programs and greater viewing pleasure. In short, cable channels have turned the absence of advertising into a tool that enhances their desirability to subscribers. Thus, pay-television, like its FTA counterpart, participates in and perpetuates consumer-centered programming. However, the product promoted by cable television is the network itself, just as a commercial film must market itself in order to attract potential viewers.

The ability for audiences to stream television programs commercial-free on host websites such as Amazon, Netflix, or Hulu further contributes to blurring the line between cinema and television. Not only does streaming significantly reduce the distance between film and television—since both can be viewed in much the same way, usually on the same website—but it also introduces new dynamics to the screening experience. In the case of television serials, it becomes possible to "binge-watch" a program, seeing as many episodes as one likes in a single sitting. Both films and television programs can now be enjoyed in one sitting and without these commercial breaks that constitute such an important factor in the relationship between audience and show on FTA television.

Yet the experiences of screening a film and viewing a television program, even when both are streamed from start to finish in one sitting, remain distinct from one another. The serialization of most conventional television programs means that viewers usually develop a far more sophisticated bond with the characters and world of a television show than those of a two-hour-long film. As Andrew B.R. Elliott has argued in the case of Arthurian serials, "unlike their cinematic counterparts which create a 'disposable' Middle Ages, serializations must invoke a believable medieval world to which they will return on a weekly basis."[10] A film director must construct a setting, atmosphere, and characters that will grasp the viewer's attention in the short span of time allotted a film preview so as to ensure a healthy audience on the film's opening weekend. In contrast, the producers of a television series can take their time describing the world they have devised. In HBO's *Game of Thrones*, for example, we are first immersed in the lives of the people of Winterfell before our attention is diverted to other important spaces (such as King's Landing). This leisurely pace is even more essential where the story's characters are concerned. Once again, a comparison with film, and especially Hollywood productions, might prove useful here. The characters of commercial film are less complex individuals capable of revealing depth, growth, and humanity than simple, generic types. Blockbuster films cannot do without the list of usual suspects: the hero, the villain, and the plucky heroine (who usually turns into a damsel in distress in the second half of the film). The hero may stray from the right path for a few minutes, but he must eventually do the right thing and return to fight the bad guy, and the villain may have endured some personal tragedy that led him to lose faith in humanity but, ultimately, he must commit some monumentally wicked deed so we, the viewers, can despise him (or, very occasionally, her). Such a formula works particularly well in epic superhero movies such as the seemingly never-ending *Spiderman* series. The serialization of television, however, makes it possible for both the greatest of heroes and the worst of villains to display some level of psychological depth. To confront viewers with the same heroic characteristics on the one hand and villainous behavior on the other would prove tedious very quickly. In order to avoid boring their viewers—causing ratings to plummet—television producers must emphasize their heroes' flaws and develop their villains' humanity to a greater extent than commercial film. The BBC's *Merlin* exemplifies the need for such ambiguity. While King Uther is portrayed as a negative character—he hunts down all those who practice magic, including children, and intimidates his own offspring, to name but two of his flaws—we also sympathize with him when Morgana, his ward and illegitimate daughter, betrays him. Likewise, we understand that Morgana feels rejected by her father and seeks revenge for Uther's oppression of magic-users. Television series like *Merlin* thus allow for the formation of a bond between characters and viewers that is much more difficult to establish when watching a film.

The personal nature of the connection between viewer and character is heightened by the intimacy with which we experience television programs. Television is ultimately a domestic medium, one that allows for audiences to enjoy their favorite shows while cooking, eating, or getting ready for bed. Unlike the public act of screening a film, which, for most people, involves dressing, some kind of transportation, and participating in a communal viewing experience with relative or complete strangers, television is usually consumed in private. For some viewers, the choice between film and television, private and public, does not even exist: "The domestic is the primary site of the audiovisual for most people at the end of the 20th century especially for, say, a single mother living in a rural setting far from a cinema."[11] Although one might argue that watching television while occupied with mundane activities lends itself to distraction rather than immersion, viewers engaging with a program at home bring its world and characters into their most private space: the show is no longer "a" show but "their" show. As a result, their involvement with television becomes even more visceral. Without others to check them from giving free flow to their emotions and to steer their response to characters and situations, they are all the more likely to project their own sympathies on what is unfolding on the screen.

What does this discussion of the main attributes of television mean for *television medievalism*? Television's consumerist bent, its serialization, and its domesticity all combine to make the medieval more comfortable and more appealing while television's democratic nature ensures that made-for-TV versions of the Middle Ages reach as many people as possible. Here, it might be useful to pause for a brief discussion of two specific aspects of this process.

First, the Middle Ages as represented on television has come to be associated increasingly with the genre of fantasy. Overall, recent television programs focusing on the Middle Ages or set against a medieval background show a marked decline in concern with historical accuracy, glorifying instead in the imaginary and the fantastical and thereby looking more like medieval romance—with its own cast of dragons, monsters, fairies, and magical adventures—than any contemporary notion of historical realism. BBC's *Merlin*, for example, does not seek, as have so many other visual retellings of the Arthurian narrative—including Bruckheimer's *King Arthur*—to place Arthur in a fixed time and place. Rather, the series unabashedly mixes the medieval with the modern—Morgana's stiletto heels—and the downright mythological, such as magic and dragons. Such lack of interest in historical "truth" is further in evidence in HBO's *Game of Thrones* set against the imaginary world of Westeros. In truth, neither of these shows attempts to depict the medieval historical era. Rather, they—like most contemporary "medieval" shows—take place in a vague pre-industrial culture, one that we are meant to recognize as proto-medieval but that is not clearly designated as such.[12]

That series like *Merlin* and *Game of Thrones* embrace and celebrate their nature as medievalisms rather than attempt to teach us about the "real" Middle Ages—one that can never truly be captured anyways—also allows for greater aesthetic license. Whereas most instances of medievalism produced during the 1980s and 1990s portrayed the medieval world as rife with injustice, disease, and misery, recent examples of television medievalism display a variety of different approaches to the subject. While some, such as Starz's *Camelot*, still revel in the idea of a grungy Middle Ages, others associate a new, "cleaner" image with the medieval period. To use *Merlin* as an example once again, the BBC production emphasizes bold primary colors, especially in the costumes of Merlin, Arthur, and the latter's knights, sweeping natural landscapes, and lavish domestic spaces.

If the clothes worn by characters in *Merlin* do not necessarily help viewers feel more at ease with the medieval, other features of this series clearly strive to solicit audience sympathy. Especially significant in *Merlin* and other recent television medievalisms is the conspicuous absence of organized religion, specifically Christianity. The Church, whose ominous presence contributed greatly to the negative image of the Middle Ages in the popular medievalism of the 1980s and 1990s, seems to have all but vanished from today's contemporary medievalisms. What is particularly striking about this phenomenon lies in its close relationship with the medium of television. HBO's *Game of Thrones* gives us some insight into the erosion of the religious on television. Although Christianity plays no role in either the television show or the series of books, *A Song of Ice and Fire*, on which it is based, the books dwell at length on the various religious beliefs of the people of Westeros. While some of this material is retained in the television version, especially in the person of Melisandre and those who, like her, worship the God of Light, much of it is pared down as if in an effort to avoid the topic of religion as much as possible. Other series either present viewers with a non–Christian world—again, *Merlin*—or one in which magic reigns (*Merlin*, *Camelot*). Without the looming "threat" of the medieval Church as it was imagined in the last quarter of the 20th century, viewers may allow themselves to form an attachment not simply with the characters of medieval-themed shows but also with the world they inhabit. Through its representation on commercial television, the medieval is thus slowly reclaiming its function as a site of popular escape and fantasy.

Television medievalism stands outside both the medieval and the contemporary, presenting audiences with the best of the two eras. Recent examples of television medievalisms eschew some prominent aspects of the medieval experience in favor of more attractive elements of the period. Such products reduce or eliminate any discussion of class, gender, and racial inequalities in the Middle Ages while emphasizing the sense of adventure and fantasy at the heart of much of medieval literature.[13] Moreover, where the Church once stood, we now find wild, untamed nature and magic. Contemporary television medievalisms offer us a romantic, pre-industrial world free of inequities and injustice, the world the Industrial Revolution led to creating without the industrialization, consumerism, and sense of awareness that played an essential part in shaping it. For the duration of our favorite medieval-themed show, we can live in the past without fearing the Church; we can feel a medieval European's sense of wonder at the world while knowing exactly where we stand in an increasingly shrinking global network. Last but certainly not least, we can pretend to ignore our society's consumerist impulses while deciding what to buy on our next outing to the supermarket. Television medievalisms thus confront us with multiple contradictions. Yet, in spite—or perhaps because—of, their paradoxical nature, such popular medievalisms serve a crucial function: they reassure us that it is possible to enjoy the comforts of modern society while continuing to grasp for some higher, more fulfilling goal.

By examining individual television programs from a wide variety of sources and an equally great range of perspectives, the ten essays in this collection further probe the nature of television medievalism in terms of Personal and Political Desires (Part 1), Narrative and Genre (Part 2), and Gender and Sexuality (Part 3). In the first essay of Part 1, "The Most Dangerous Sport in History Is About to Be Reborn: Medievalism and Violence in *Full Metal Jousting*," Angela Jane Weisl discusses how the History Channel's *Full Metal Jousting* (2012) enacts a simultaneous engagement with and rejection of the Middle Ages, creating an "authentic fantasy" of the Middle Ages that is fraught with the tensions and ironies of postmodern masculinity. Stephanie L. Coker's piece, "*Joan of Arcadia:* A Modern Maiden on Trial," explores what it means for Jeanne d'Arc to find her place in American pop culture by playing a role in civic education in a television series aimed at teenagers. This essay provides a close reading of episode nine, "St. Joan," of *Joan of Arcadia* (2003–05) wherein high schooler Joan Girardi reconciles religious faith and secular education. Coker argues for how *Joan of Arcadia* refuses the spectacle in Jeanne's story in favor of confronting the "mysteries of faith in a non-religious age." In "William Webbe's Wench: Henry VIII, History and Popular Culture," Shannon McSheffrey turns to an analysis of one of the more obscure scenarios from Henry's sex life included in the BBC/Showtime *Tudors* (2007–10). This exploration reveals much about what kind of cultural work television does for us, and indeed about how the popular culture of the past underlies our scholarly work. Such cultural work is also the focus of Evan Torner in "Nature and Adventure in *Die Jagd nach dem Schatz der Nibelungen*." This fourth essay examines how the made-for-TV adventure film is a very serious national-pedagogical project attempting to re-enchant what is seen as a shared German medieval past while glossing over the recent and far more disturbing history of the Third Reich.

In Part 2, authors turn to issues of narrative and genre while attending to medium, audience, and context. Melissa Ridley Elmes's "Episodic Arthur: *Merlin*, *Camelot* and the Visual Modernization of the Medieval Literary Romance Tradition" suggests how useful it is to consider television programs such as the BBC's *Merlin* (2008–12) and Starz's *Camelot* (2011), rather than their film counterparts, as "the modern visual successors of the medieval literary romance

Introduction

Television Medievalisms

MERIEM PAGÈS and
KAROLYN KINANE

The past decade and a half has seen an impressive resurgence of popular interest in the Middle Ages. Since 1997, J.K. Rowling's acclaimed *Harry Potter* series and Peter Jackson's cinematic adaptation of Tolkien's *Lord of the Rings* trilogy (2001–03) have led to a renewed exploration of the medieval past—or at least our magical and mythical fantasies about it. Today, such excitement for all things medieval continues unabated, especially in small screen medievalisms. Television presents us with a wide and diverse array of "medieval" offerings such as BBC's *Merlin* (2008–12), HBO's *Game of Thrones* (2011–), and Starz's *Camelot* (2011), among others. The last few years also bear witness to a slow, yet noticeable shift in the type of medium associated with medieval-themed narratives. Whereas the Middle Ages used to inhabit primarily the world of film—where sweeping natural landscapes and scenes of carnage could be emphasized in all their glory—depictions of the medieval have come to fit more and more comfortably on the small screen. Most of the last major film productions to deal with the Middle Ages, such as *A Knight's Tale* (2001), *King Arthur* (2004), and *Beowulf* (2007), date to the first decade of the 21st century while television programs set against a medieval background continue to attract viewers well into 2014.[1]

Despite the growing number of medieval-themed series on commercial television on the one hand and the importance of the connection between popular medievalism and the medium of television on the other, there exists at present very little scholarship on the image of the Middle Ages in television. Unlike tradition," highlighting the medievalism inherent in the very notion of "episode" and episodic storytelling. Both shows are anachronistic and ahistorical, presenting the Arthurian legend in ways calculated to appeal visually to a modern audience steeped in the enchantment of medievalism. In this, Elmes argues, they are very like their medieval counterparts that also sought to inculcate a sense of timeless nostalgia. Complementing her colleagues' work on adult programs, Sandy Feinstein's "Are You Kidding? *King Arthur and the Knights of Justice*" explores how the sport-metaphor in animation series such as *King Arthur and the Knights of Justice* (1992–93) shapes the political, specifically royal, medieval past for juvenile audiences. By situating this animated serial within a history of remediation, with particular attention to the legacy of Mark Twain's *A Connecticut Yankee in King Arthur's Court*, Feinstein further theorizes the notions of fandom, technology, and immediacy.

The final four essays are concerned primarily with gender and sexuality in television medievalism (Part 3). In "Television's Male Gaze: The Male Perspective in TNT's *Mists of Avalon*," Michael W. George places this adaptation in the context of televised "warrior women" and "strong female characters" such as *Buffy the Vampire Slayer* (1997–2003) and *Xena Warrior Princess* (1995–2001) to demonstrate how the 2001 TNT miniseries *Mists of Avalon* reinforces the male gaze. By analyzing characterizations, camera angles, zooms, and crosscuts, as well as features of Marion Zimmer Bradley's novel that Edel omitted, this essay demonstrates that the miniseries "androcentrizes" Bradley's narrative, "bringing it into line with not only much of Arthuriana but also much of television medievalism." The androcentric nature of many instances of television medievalism is further emphasized by Elysse T. Meredith in "Gendering Morals, Magic and Medievalism in the BBC's *Merlin*," an essay that explores the gradual othering of female magic in the series. Meredith suggests that *Merlin* is a medievalism of gendered morality that comments on religion, monarchy, and the role of governmental devolution in contemporary British society. This essay further demonstrates how contemporary Arthuriana can be potently nationalistic, serving as quasi-propagandistic material that expresses anxieties over contemporaneous political and social issues by depicting corresponding subjects as inherently dangerous.

The volume's penultimate essay, Tara Foster's "*Ne cherchez pas la femme*: The Women of *Kaamelott*," allows for a more optimistic reading of the intersection between gender and medievalism on the small screen. Using Maureen Fries's categorization of Arthurian women—heroine, female hero or counterhero—as a framework, Foster demonstrates how the serial format allows a character such as Guenièvre to move among various roles. Foster suggests that Guenièvre and other female characters not only contribute to the sexual humor of the French series *Kaamelott* (2005–09) but also interrogate medieval and modern gender inequities, indicating the extent to which the Arthurian legend still serves as a powerful framework for evoking societal ideals. Finally, Torben R. Gebhardt's "Homosexuality in Television Medievalism" explores how homosexuality is represented in HBO's *Game of Thrones* (2011–).

Each of the essays that follow testifies to the importance of television in fashioning a "new" Middle Ages for the 21st century. That medieval-themed narratives seem to be migrating more and more from the big to the small screen might initially appear to reflect a gradual loss of interest in the medieval. However, this is far from the case, and the popularity of the medieval period on television—as proven by the abundance of programs set against a medieval background—only helps to draw attention to the ubiquity of medievalism. At a time when cinema has begun to wane as a popular medium, especially for young adults, television is providing medievalism the opportunity to grow. Through television, popular images of the Middle Ages retain their currency, adapting to the needs and desires of various new audiences and ensuring that the medieval can continue to capture the hearts and minds of succeeding generations.

Notes

1. Ridley Scott's *Kingdom of Heaven* (2005) and *Robin Hood* (2010) present two interesting exceptions to this rule.
2. The popularity of video games such as *Assassin's Creed*, *Crusader King II*, *Mount and Blade: Warband*, and *Chivalry: Medieval Warfare* stress the importance of this new form of medievalism.
3. Ute Berns and Andrew James Johnston, "Medievalism: A Very Short Introduction," *European Journal of English Studies* 15.2 (2011): 97.
4. John M. Ganim, "The Myth of Medieval Romance," in *Medievalism and the Modernist Temper*, ed. R. Howard Bloch and Stephen G. Nichols (Baltimore: Johns Hopkins University Press, 1995), 148.
5. Ganim, "The Myth of Medieval Romance," 148. Of course, not all 19th-century medieval enthusiasts embraced Morris's leftist ideology. Here, it is important to note that the Middle Ages inspired individuals from a wide political spectrum from the eighteenth century onwards.
6. Richard Utz, "Robin Hood, Frenched," in *Medieval Afterlives in Popular Culture*, ed. Gail Ashton and Daniel T. Kline (New York: Palgrave Macmillan, 2012), 153.
7. Steve Guthrie, "Time Travel, Pulp Fictions, and Changing Attitudes Toward the Middle Ages: Why You Can't Get Renaissance on Somebody's Ass," in *Medieval Afterlives in Popular Culture*, ed. Gail Ashton and Daniel T. Kline (New York: Palgrave Macmillan, 2012), 110.
8. Guthrie, "Time Travel, Pulp Fictions, and Changing Attitudes toward the Middle Ages," 109.
9. Angela Jane Weisl, *The Persistence of Medievalism: Narrative Adventures in Contemporary Culture* (New York: Palgrave Macmillan, 2003), 204–6.
10. Andrew B.R. Elliott, "The Charm of the (Re)making: Problems of Arthurian Television Serialization," *Arthuriana* 21.4 (2011): 53.
11. Rod Stoneman, "Nine Notes on Cinema and Television," in *Big Picture, Small Screen: The Relations between Film and Television*, ed. Elizabeth Sklar and Donald Hoffman (Jefferson, NC: McFarland, 2002), 120.
12. Based on medieval Norse sagas, the History Channel's *Vikings* (2013–) presents an interesting exception to this rule.

13. ABC's *The Quest* (2014–), for example, represents a new fantasy/reality television hybrid that gives its contestants the opportunity to populate a fantasy medieval world void of racial and gender discrimination.

Bibliography

Aberth, John. *A Knight at the Movies: Medieval History on Film*. New York: Routledge, 2003.

Ashton, Gail, and Daniel T. Kline, eds. *Medieval Afterlives in Popular Culture*. New York: Palgrave Macmillan, 2012.

Berns, Ute, and Andrew James Johnston. "Medievalism: A Very Short Introduction." *European Journal of English Studies* 15.2 (2011): 97–100.

Clements, Pamela, and Carol L. Robinson, eds. *Neomedievalism in the Media: Essays on Film, Television, and Electronic Games*. Lewiston, NY: Edwin Mellen, 2012.

Driver, Martha W., and Sid Ray, eds. *The Medieval Hero on Screen: Representations from Beowulf to Buffy*. Jefferson, NC: McFarland, 2004.

Elliott, Andrew B.R. "The Charm of the (Re)making: Problems of Arthurian Television Serialization." *Arthuriana* 21.4 (2011): 53–67.

Ganim, John M. "The Myth of Medieval Romance." In *Medievalism and the Modernist Temper*, edited by R. Howard Bloch and Stephen G. Nichols, 148–66. Baltimore: Johns Hopkins University Press, 1995.

Harty, Kevin J. *The Reel Middle Ages: American, Western and Eastern European, Middle Eastern and Asian Film about Medieval Europe*. Jefferson, NC: McFarland, 2006.

Morgan, Gwendolyn. "Gnosticism, the Middle Ages, and the Search for Responsibility: Immortals in Popular Fiction." In *Medievalism in the Modern World: Essays in Honour of Leslie J. Workman*, edited by R. Utz and T. Shippey, 317–27. Turnhout: Brepols, 1998.

Stoneman, Rod. "Nine Notes on Cinema and Television." In *Big Picture, Small Screen: The Relations between Film and Television*, edited by Elizabeth Sklar and Donald Hoffman, 118–32. Jefferson, NC: McFarland, 2002.

Weisl, Angela Jane. *The Persistence of Medievalism: Narrative Adventures in Contemporary Culture*. New York: Palgrave Macmillan, 2003.

Part 1

Personal and Political Desires

The Most Dangerous Sport in History Is About to Be Reborn

Medievalism and Violence in Full Metal Jousting

Angela Jane Weisl

Difficult as it might be to believe, jousting is enjoying a revival because, at least in part, "there's a real possibility of getting hurt," as a spectator at the Gulf Coast International Jousting Championships noted in anticipation.[1] Here, the Middle Ages represent the possibility of authorized violence and the spectacle of witnessing its combatants injuring one another. In the winter of 2011–2012, jousting made its way to American television in two shows, the National Geographic Channel's *Knights of Mayhem*,[2] featuring champion jouster Charlie Andrews and his troupe preparing for the West Coast Jousting Championships, and the History Channel's much more successful *Full Metal Jousting*.[3] While the former show, despite offering plenty of violence, was essentially the *Real Housewives of Jousting*, presenting somewhat unappealing characters shouting profanities at each other, *Full Metal Jousting* attempted to bridge a gap between the jousting performances audience members might have seen at Renaissance Fairs and Medieval Times™ and the contemporary sport, showing both the pleasures and challenges of taking on an ancient practice in the modern world.

That said, the gleeful spectacle of destruction is exhibited in *Full Metal Jousting,* which featured jousters spitting up blood, getting injured, and falling to the ground in the lists. In this reality show, hosted by champion jouster Shane Adams, sixteen contestants with varying degrees of relevant experience (while roughly half came from the world of theatrical jousting, the others all had significant horsemanship experience, either as rodeo riders, polo players, horse trainers, show jumpers, or three-day eventers) were taught how to joust and then pitted against each other in a tournament with a $100,000 purse. The contestants were divided into two teams, Red and Black, with each team coached by a highly-ranked jouster: Red by American Ripper Moore, and Black by Australian Rod Walker. The team that controlled the field (whichever one had won the previous joust) would choose a member of its team to joust and select his opponent from the other side. After scenes of training, the joust would take place. After the eight preliminary jousts, the victors would be paired by the coaches into four quarter-finals; those winners would face off against each other in the semi-finals, with the two victors then jousting for the $100,000 prize. Because of the dangers of jousting, no contestants were sent home; losers had the chance of returning to the lists should anyone become unable to compete. Two competitors did indeed return for second preliminary jousts. Halfway through the series, a consolation final was announced, although unlike most bronze metal matches, this would not be between the two losers of the semi-finals, but between jousters chosen by each team, with a prize of $25,000. Whether this surprise was instituted from the beginning or designed to make up for contests not turning out as the producers had hoped was unclear, but it did help to keep enthusiasm and morale up among the eliminated jousters. Although *Full Metal Jousting* followed many of the conventions of reality television—the contestants all lived together in one house and were divided into two teams, each with a set of coaches—*Full Metal Jousting* focused more on the training and jousting than it dwelled on interpersonal relationships and conflicts; nevertheless, it was often through the presentation (and self-presentation) of the various participants (rather than the jousting itself) that the show's medievalizing elements came most strongly to the fore.

The show itself enacts an odd engagement with and rejection of the Middle Ages. The presentation of the sport itself put forward an interesting tension between jousting's historical elements and its modern ones, although the voice-over narration often represented medieval elements as modern and vice versa. This tension is encapsulated in the armor the contestants wore, specially made for the show by Anshelm Arms, which made the jousters look like a cross between medieval knights and Iron Man. With their names on the back, the relationship to contemporary sports was emphasized; yet throughout the program, parts of the armor—the gridded grand guards, gauntlets, gambesons, and tassets—were referred to with authentically medieval names, often without any explanation of what they were. Indeed, Shane Adams described the armor as having "all the elements of history with a modern touch."[4] While many of the training apparati, such as the tilting field and spinning and shot quintains were quite authentic, having contestants being hit with baseball bats to simulate getting hit with a lance seemed far more modern.[5] Using a battering ram for the same purpose created a rather ironic nexus between past and present. The tournament rules, which awarded points for direct hits, shattered lances, and unhorsings, actually reflected late medieval practices, despite Shane Adams' rather erroneous claim that "in history the emphasis was on breaking a lance, but in this version, the biggest points are from unhorsings. That's the American touch."[6] It seemed throughout the series that the show was attempting to use violence to distinguish contemporary jousting from its medieval antecedent; ex–Marine and MMA fighter David Prewett commented, "I never wanted to be a knight, I've never dressed up as a knight, I've never been to a show ... here for the adrenaline rush,"[7] while cowboy Nathan Klassen said, "I'm not here because I love medieval times. I'm here to be competitive, to hit people."[8] In an interview with Scott Maslow of the *Atlantic,* Shane Adams claimed, "back in the 15th and 16th centuries, most of the competitors jousting were nobility. So when they were out jousting, they were putting on more of their own theatrical display. Of course, they went out to win, but the lances they were using were not war lances, and were made to break, because these nobles didn't want to go out there killing each other,"[9] implying that contemporary jousting has a greater commitment to violence and destruction than its medieval antecedent. However, reveling in violence is hardly specific to contemporary jousting; Richard Kaeuper notes of the literature of chivalry and tournament that "almost without fail these works give prominence to acts of disruptive violence and problems of control,"[10] and David Crouch points out that "in a description of the Jousting à Chauvency in January 1280, poet Jacques Bretel 'gives detailed descriptions of the jousts, the heralds' commentaries, the injuries and the reactions of the crowd. When one pair were badly bloodied, the ladies commented how the jousting that day was 'impressive and fine,' and when two lances simultaneously burst into fragments that 'it was jolly well done.'"[11] The tournament Chrétien de Troyes describes in *Erec and Enide* is hardly devoid of violence:

> And the knights
> Were ready. The noise was tremendous,
> Metal crashing on metal:
> Lances shattered, shields
> Cracked, mail shirts split.
> Saddles were emptied, as knights
> Went tumbling. Horses sweated
> And foamed [2112–2118].[12]

Indeed, one of the show's most peculiar misreadings of the Middle Ages ties in to the emphasis on jousting's status as history's "Most Extreme Sport," as it is well documented that jousting outpaced the *mêlée* in popularity as the go-to tournament activity because, as an individual combat, there were fewer injuries and deaths, making it, at best, history's slightly less extreme sport.[13]

This attempted distinction often turned on itself in the show; to make full-contact jousting "the real thing," it had to distinguish itself from theatrical jousting, performed at Renaissance Fairs and Medieval Times,™ from which many of the contestants, including three of the top four finishers, came. In the opening episode, for instance, Shane Adams announced, "We've all seen dinner show jousting, and this is not that," and Matt Hiltman noted, "I thought I was really tough doing theatrical jousting. I wanted to see what the real deal was like."[14] In an interview, Shane Adams added, "The only way you're going to make it a real sport is by doing the real thing. We're putting together and producing a modern-day venue for a modern-day sport. But we're still using a lot of historical elements to bring it there."[15] If "real" here is opposed to "medievalism," which is presented as inauthentic, staged, a spectacle, is "real" therefore the medieval or the modern? Is *Full Metal Jousting* real because it is *more* medieval than its theatrical counterparts, or because it places its own medievalism as somehow separate from the elements of pageantry associated with the medievalism of the Renaissance fair? Certainly, full-contact jousting is more physical than both the theatrical productions and the historical reenactments popular in Europe; however, does that quality of presence—of actual hits, violent unhorsings, and nasty injuries—really reclaim a truly medieval experience? And how much does an authentic medieval experience matter to these modern knights? Is the claim to authenticity challenged by the deeply theatrical elements of medieval jousting itself? Stripping away the pageantry essential to the medieval tournament, and calling the contact and combat the "real thing," beg an element of the question; in the Middle Ages, of course, jousting and tournaments were emphatically not "real" no matter how hard the hits or how bad the injuries; the "real thing" was armed combat for which tournaments served as a kind of preparation in times of peace. Here the show's uneasy relationship with the medieval past comes to the fore; while it certainly acknowledges that any jousting performed in the 21st century is not truly medieval—after all, the Middle Ages cannot be reclaimed in any truly authentic way, no matter how hard the contestants hit each other with their lances—it also seems uncomfortable in its association with what the sport has come to be seen as—a performance that abandons both violence and chivalry in favor of a neo–Arthurian fantasy. In a strange way, contemporary jousting, exemplified by *Full Metal Jousting*, enacts its own medievalism through a peculiar set of longings—simultaneously for a past stripped of certain kinds of pageantry and performance and a past in which violence and danger are valorized through a set of ritualized values.

The medievalism that the show rejects as "unreal" also has more authentic roots than it probably gives itself credit for, as Steven Jeffreys points out:

> All these romantic ideas (courtly love) began to change the idea of the tournament. A totally new kind of tournament evolved, called a "Round Table"; this began with the competitors taking strict vows to behave honorably and to obey the laws of chivalry, as King Arthur and his companions were thought to have done. Ladies were given seats of honour in the stands to watch the noble proceedings."[16]

Jousting's and tournaments' roles in medieval courtly love, real and literary, is extensive, yet the show sought to push all elements of femininity to the margins in favor of the aforementioned violent, yet chivalric, masculinity. One doesn't want to suggest any knowledge of medieval tournament history on the part of the producers of the show, but it was certainly notable that there were no women to be seen anywhere in *Full Metal Jousting*; even all the horses were male. The courtly love elements so important in the later medieval tournament and most of its literary examples appeared only in some of the competitors' references to wives and daughters for whom they fought, and these were certainly never presented as the focus of their efforts. Once again, one element of the Middle Ages was denied in favor of a competing one. While medieval literature, and to an extent, medieval life made no separation between love and violence (courtly love was explicitly constructed for the knightly classes), *Full Metal Jousting* insisted on separating one medievalism from another, authorizing it, and then claiming this fiction of medieval masculinity to be somehow more authentic than any other.

If conventions of courtesy played no role in the show, the show nevertheless put forward medieval chivalric ideals almost despite itself; bravery, valor, determination, and a kind of selflessness (provided these didn't affect the jouster's performance in the lists) were highly valued, while selfishness and inappropriate behavior were condemned. Indeed, Shane Adams claimed repeatedly that he always wanted to be a knight, and as the world's top-ranked professional jouster, also performs as part of a troupe called the Knights of Valor, whose motto is "Bravery, Honor, Chivalry." This motto may well have provided some of the motivation for these ideals' appearance on the show, as his whole engagement with the sport of jousting suggests an affinity for medieval values. Whether intentional or unintentional, *Full Metal Jousting* offered up a set of values very much in line with the chivalric code, or at least the chivalric code as it is contemporarily imagined. This is not to suggest a separation of these values from the violence with which the show sold itself, but rather a kind of integration of them; in describing medieval chivalry, Kaeuper observes that "however glorious and refined its literature, however elevated its ideals, however enduring its link with Western ideas of gentlemanliness—and whatever we think of that—we must not forget that knighthood was nourished on aggressive impulses, that it existed to use its shining armour and sharp-edged weaponry in acts of show and bloody violence."[17] *Full Metal Jousting* might have expressed its own chivalric values in an inverse way; however much the show was about hard hits, pain, combat, and aggression, the audience could not forget that being a knight meant adhering to a set of ideals, not exactly medieval, but not exactly separated from those that defined the original chivalry either. Certain attitudes and behaviors were shown to be more positive—and ultimately led to more success—than others.

In the show's most overt engagement with chivalric behavior, Landon Morris, an auctioneer from Australia who had barely been featured up to this point in the series, was sent home for mistreating his horse. In Episode 5, at Black Team practice, his horse stepped on him, and to make it move, he punched the horse in the head with his gauntlet on. The coaches' action was punitive and swift; Landon got a dressing down from Shane, Rod, and Ripper and was immediately sent packing. The coaches' comments were themselves revealing; Shane commented, "The horses are here to be your teammates, not to be mistreated," adding to Landon, "You have a lot of growing up to do." Landon's insufficient remorse was met by Rod saying, "Step away and keep your mouth shut"; he also commented, "The fact that Landon mistreated a horse—amazingly arrogant, one of the stupidest things I've ever seen."[18] Here he echoes the sentiment of Geoffroi de Charnay, in his *Book of Chivalry:* "Where humility is to be found, there reigns good sense and happiness" (133).[19] Raymond Lull, author of a key medieval guide to chivalry, informs us, "It was found that the horse was the most noble and most convenable to serve man, and because of that, among all beasts, men chose the horse and gave him to this same man. For after the horse which is called cheval in French that the man named chevalier which is knight in English. Thus the most noble man is given the most noble beast."[20] Mistreating a horse was therefore tantamount to mistreating another knight; failing to treat the horse with nobility was to abrogate one's own nobility. The overall sentiment was that Landon did not know his place and deserved his punishment; Josh Knowles' remark "Landon made a mistake, and he had to pay heartily for it" was taken one step further by Rope Meyers, whose comment "Landon violated the conduct rules of this competition, and they had to do what was right"[21] showed the intimate tie of ideology to action. Outside viewers concurred; Linda McCabe, in her review of the show, noted, "The man did not demonstrate respect for his horse and therefore has no honor."[22] As a knight (even a 21st-century one), there are codes of chivalric conduct that must be observed; Landon failing to do so meant he was unfit to be a jouster, regardless of what his skills with the lance might have been. If abusing his horse violated one aspect, his disrespect for the coaches' decision violated another; violence against animals and insufficient humility in the face of one's superiors express values simultaneously medieval and modern.

Some of these values included a willingness to withstand pain and injury, seriousness of purpose and focus, competitiveness, a desire to better one's life and the life of one's family by means of the competition, and a commitment to jousting as an embodiment of a set of ideals. Competitors who came primarily to win the purse were presented as problematic. Josh Avery, one of the show's most accomplished jousters, praised for his skill and accuracy, could not achieve approbation because he repeatedly said, "I came here to win the money"[23] and seemed a bit arrogant ("I'm definitely the most accurate guy here"; "I was breaking lances whenever I wanted to break them"[24]) about his abilities and later rather grousy. When discussing his chances against Josh Knowles in the semifinal, he commented, "He's bigger than me, he's stronger than me; he likes to remind me of that all the time. But I'm just as mean," casting himself as the black knight, despite coming from the Red Team (and, ironically, despite Josh Knowles playing the part of the evil knight in the Medieval Times™ show from which they both came).[25] Making his claim to participate in the consolation joust by saying, "I want $25,000" meant he lost in the voting to John Stykes who put forward the more chivalric "I don't care about the money, I just want to joust."[26] This isn't to say that a desire for the prize was inappropriate; Kaeuper comments of medieval knights that "of course acquiring land and wealth is assumed to follow naturally, and is welcomed as an enhancement of honour."[27] However, to make the acquisition an enhancement of honor on *Full Metal Jousting*, it seems it had to be undertaken for a higher purpose, the desire to provide for home and family, or simply as a testament to a successful performance.

Mike Edwards, who at first seemed a likely contender for the title, was criticized for surliness and sulking after losing his first joust to Josh Knowles, leading Rope to comment that Mike "has become really abrasive."[28] Mike's suggestion that he should go home was met by Shane's scorn: "Mike was a leader for the Red Team, and he failed. But by giving up, he's an even bigger failure. Clear your mind, concentrate on training, and become the best you can be." Mike's response to Shane's dressing down expressed his attachment to the *comitatus*: "I'll stay around for you."[29] Mike was given a chance to joust again by being moved to the Black Team after Brian Tulk's training injury took him out of the competition, but could not redeem himself; the Black Team was somewhat reluctant to embrace him, and when he was announced in the lists in the joust against Nathan Klassen, no one was seen to applaud. When he lost 17 to 3, Jack Mathis, another somewhat ambiguous figure who wanted the opportunity to joust a second time himself, commented, "Now the whole world knows they should have picked someone else."[30]

If Josh Avery, Jack Mathis, and Mike Edwards vied for the position of the antagonistic Sir Kay, several other figures embodied the show's notions of chivalric virtues. James Fairclough, called "the Hammer," exhibited talent, fortitude, and strength despite being the smallest jouster and coming from the world of show jumping, an incredibly dangerous and skilled sport that is rather ironically viewed as being somewhat effete. Being able to control the most difficult horse and coming back from a concussion and head wound that required several staples to hold it together gave James a reputation for excellence. Matt Hiltman, billed as a Philosophy Student, which was clearly supposed to suggest a somewhat contemplative and perhaps overly cerebral nature, although he was also a skilled theatrical jouster, demonstrated an astonishing ability to withstand major injuries and painful unhorsings. Paul Suda, despite expressing a (quite reasonable) fear of injury, let go of his anxieties, won a decisive victory over heavily-favored polo player Tom Conant, and proceeded to put up a good showing against Josh Knowles in the quarterfinals.

Yet the show's most medievalizing move was the near canonization of Rope Meyers, a steer wrestler from Texas, who was chosen first in the pick, and from that point forward emerged as the show's Sir Gawain (of *Sir Gawain and the Green Knight*), the "perfect Christian knight," praised for both his prowess and his chivalry. Rope, while eager to win and in touch with his competitive side, became the voice of good values, good sportsmanship, and honor. He seemed, for instance, to have internalized Geoffroi de Charny's maxim "Speak of the achievements of others but not of your own, and do not be envious of others."[31] Always quick to praise his opponents and compatriots, when jousting with Joe McKinley in practice, he commented, "Joe was doing phenomenally well. He just kept getting better and better, more comfortable on his horse. He's going to be a force to be reckoned with in this competition."[32] Ironically, Rope's first joust came in the same episode that saw Landon sent home; at the pick, he was chosen to take on David Prewett, an ex–Marine, MMA fighter, and rodeo rider. Downplaying his own abilities, Rope would only say, "If I have any advantage, it's in my riding." After a terrible practice, he commented, "You know humility is a good thing." However, he also added, "I never let winning be the thing that ultimately guides me." While David viewed the joust as a chance to take out his anger and aggression over the death of family members and fellow marines, Rope seemed to have higher aspirations. Before the joust, in what appeared to be a scene taken from a medieval (or perhaps pre–Raphaelite) painting, Rope was pictured in full armor, kneeling in a shaft of light, praying:

> Thank you, Father, for this opportunity, thank you for the chance to do this ... I'll be fearless, courageous, honorable, and above all else, Father, that my kids will see this and know that their Dad loves them and that they—he's trying to live as telling them to live. I thank for that in Jesus' name. Amen.[33]

After his victory, he thanked his teammates, his family, and even Praetorian, his horse. Clearly, he has taken a page out of Geoffroi de Charny's book:

> You should, therefore, always and in all circumstances be determined to do your best, and above all have the true and certain hope that comes from God that He will help you, not relying just on your strength or your intelligence nor your power but on God alone, for one often sees that the best men are defeated by lesser men, ... You can see clearly and understand that you on your own can achieve nothing except what God grants to you. And does not God confer great honor when He allows you of His mercy to defeat your enemies without harm to yourself?[34]

In the semi-final joust, Rope competed against Josh Avery, the aforementioned antagonistic figure, known as "wolfman" for his propensity to howl, and the most overt of the show's successful fighters in his claim that he was primarily there to win the $100,000 ("I came for the money. This is probably the only opportunity I'll ever have to get a 6 figure paycheck"). Before the competition Rope praised his opponent ("I think Avery is a stud") and reprised his Joan of Arc scene, praying "I thank you Father for this awesome opportunity and I pray that I will do my very best with it. Do all of it. As unto you Father. And, uh, Father, thank you. This is awful fun. In Jesus' name. Amen."[35] In one of the best jousts of the competition, Rope was finally bested 19 to 18, and again displayed humility in the face of defeat, embodying another of Geoffroi's exhortations: "And if you are defeated, does not God show you great mercy if you are taken prisoner honorably, praised by friends and enemies."[36]

All these qualities led Rope to be the Black Team's selection for the consolation round joust in a somewhat overdetermined drama; after Joe declared (for no particular reason that the audience was privy to) that "Rope and Jack are the only ones eligible,"[37] the team ultimately left the decision to a coin toss which went in Rope's favor. Viewers familiar with medieval romance's affinity for coincidence might well have been tempted to declare, "Of course it did," and repeat "Of course he did" at Rope's final victory. Entering the joust declaring, "Going into this match up it's not about the competition itself. I really thought about my kids and my family and what kind of legacy I want to leave for them," Rope managed a 6 to 2 win over John Stykes in a fairly sloppy and uneventful contest. Instead of celebrating, however, he had his most Gawain-like moment, apologizing for his poor performance, and once again expressing humility. And like Gawain, he feels himself changed; in his declaration, "I came into this competition thinking it would be a very interesting thing to try and be a part of and I feel like I've left this competition a jouster, and now that's part of who I am and what I am. *Full Metal Jousting* is part of my bloodstream,"[38] viewers witness his sense of transformation through this chivalric experience. His blog, "Crosstraining Journals," explores these sentiments more fully, exploring the nature and value of humility, honor, his admiration for his fellow jousters, and the connection of all of these things to God in entries called "Humility," "Courage," and "Winning Ugly,"[39] showing that his commitment to chivalric values is not simply imposed by the show's producers, but authentic and deeply felt.

Ultimately, the show reflected medieval knightly values in its emphasis on prowess. It seems unlikely that Rope Meyers would have achieved the same level of adulation had he not been such a formidable opponent in all but his final joust. Drawing on medieval antecedents, prowess was not simply brute force but a combination of physical power, riding ability, and accuracy with the lance. Kaeuper notes, "The key trait of knightly prowess wins divine approbation; disloyalty and anything leading to dishonor becomes sin, a moral and not merely a social blunder"[40]; this can sum up why Rope triumphed and Landon was sent home. But as Kaeuper adds, "prowess was truly the demi-god in the quasi-religion of chivalric honour; knights were indeed the privileged practitioners of violence in their society."[41] In selling the show with the tag line "The Most Dangerous Sport in History Is About to Be Reborn," it is clear that *Full Metal Jousting* embodied this medieval conjunction. Over scenes of tough hits and unhorsings, the voice over continued, "Sixteen of the toughest warriors in the world have been chosen. They'll ride 2000 pound war horses. They'll have 80 pounds of steel on their backs. They'll face off in the ultimate test of skill and guts. Only one will survive to win the $100,000 and become the first champion of *Full Metal Jousting.*"[42] The show offered two intersecting definitions of prowess, success in the joust and the ability to take hits. The show clearly valued the culture of "playing hurt"; each episode displayed contestants choosing to continue their jousts despite injuries, while those who surrendered to their wounds (and thus their foes) often expressed feelings of embarrassment or shame, or, if they didn't, were castigated for their failure to do so. When Brian Tulk pulled out with a groin pull, Shane expressed significant doubts about his toughness: "I would joust with a broken leg. I would joust with a muscle tear. I would joust with my leg half missing. I would. It's 100,000 dollars."[43] Matt Hiltman's response to his own gruesome injury (a major gash in his upper thigh requiring several stitches—a rather medieval wound in itself) embodied Shane's definition of fortitude—"The only risk I'm running is reopening it, and that's a risk I'm willing to take"[44]—as did James Fairclough's return to riding as soon as he was cleared after his concussion. "For the medieval knight, the risks were worthwhile. Jousting offered an unparalleled opportunity for a knight to display his prowess before audiences of as many as 10,000 spectators and win acclaim, horses and mercenary equipment," Dashka Slater notes.[45] Failure to compete, even due to injury, often results in expressions of self-doubt and humiliation, a sense of failure comparable to Sir Gawain's sense of his abrogation of his chivalric responsibility at the end of *Sir Gawain and the Green Knight*, despite his success in the martial elements of his quest. This is not without medieval antecedent either; too much attention to the body, Geoffroi of Charny notes, should be avoided: "Men of worth tell you that you must in no way indulge in too great fondness for pampering your body, for love of that is the worst kind of love there is. But instead direct your love toward the preservation of your soul and your honor, which lasts longer than does the body, which dies just as soon, wither it be fat or lean. Too great a desire to cosset the body is against all good."[46] So, too, does failure of prowess; for example, Mike Edwards' disgust at his own inability to achieve the goals he desired despite two opportunities, for as Kaeuper points out "characters who have been defeated in the initial, mounted fight with lances, often declare that they have been 'shamed.'"[47]

The show's investment in the values of aggression, withstanding of pain, and conquering fear raise a larger question, however. In her *New York Times* article which might well have brought jousting to the attention of network executives, making them willing to consider proposals for both jousting shows, Slater comments that "lurking underneath the surface of the debate over jousting styles are deeper questions about masculinity itself."[48] Fantasies of gender flourish in "medieval" experiences, often pining back to stereotypes of chivalrous knights and their maidens fair, yet such gender roles demand rigorous adherence to their traditional parameters. Indeed, as Clare Lees notes of medieval masculinity, "the burden of masculine potency (symbolic or real), shadowed by impotence, exacts a heavy price,"[49] an observation applicable as well to many masculinities of today. The nexus of violence, masculinity, potency, and 11-foot long lances nearly provides a "write your own joke" moment, yet that too, has its medieval predecessors; discussing medieval Iberian jousting, Noel Fallows notes, "From both a physical and a symbolic point of view jousts offered tangible evidence of a man's prowess, of the meaningful role that he played in the masculine active life and of his rejection of a life of sloth or recreance.... In what has to be the most erotic jousting metaphor ever conceived, the fictitious knight Tirant Lo Blanc's prowess in the enclosure is compared to his prowess in the bedroom."[50] While the show never resorted to schoolboy jokes, it was clear that he who wielded the most accurate and powerful lance was also the most masculine. It is perhaps over-reading to understand this particular return to medievalism as an attempt to re-inscribe traditional notions of gender against the anxieties of a shifting modern world, but it is notable that while women (wives, girlfriends, daughters) are mentioned, none appear on *Full Metal Jousting*. Like the lady of medieval romance, they are essentially theoretical, an inspirational fantasy that may make the jouster try harder but doesn't interfere in his participation in an entirely-masculine world. Without women, it seems, there is no challenge to aggressive behaviors (no one suggesting that the jouster should stop risking life and limb and instead get a real job), other than those imposed by the show's notions of chivalry. The program offers the fully masculine world of the *comitatus*, defined both by loyalty (the teams cheer on and support each other in success and failure) and by competition. After winning his joust against the heavily-favored Jack Mathis, Jake Nodar, the show's "eccentric horse trainer,"[51] instead of congratulating himself, expressed the essential nature of this bonding: "the feeling of having joust control back to the Red Team is awesome. The team has been behind me 100% getting ready for this, and being able to do this for them is huge."[52] Despite team loyalty, friendship was never an impediment to competition; on several occasions, the two jousters revealed various kinds of connections: Josh Knowles and Josh Avery worked together; Josh Knowles had trained Paul Suda in theatrical jousting; Jake Nodar had to joust teammate James Fairclough in the quarter-finals, yet all these contests were fiercely competitive. Like medieval knights before them, "they fought each other as enthusiastically as any common foe."[53]

The show encodes a desire for raw virility and aggression that cannot be divorced from performances of gender transposed across the centuries. Certainly violence is a characteristic of masculinity, medieval or modern, and as Ruth Mazo Karras observes,

> Violence as a central feature of masculinity is, of course, by no means limited to medieval aristocracy. Pierre Bourdieu is hardly alone when he generalizes that "manliness must be validated by other men, in its reality as actual or potential violence, and certified by recognition of membership in the group of 'real men.'" It is much more prominent, however, in some historical contexts than others. In the late Middle Ages, violence was the mode of masculine expression within knighthood [21].[54]

For this reason, the comments of the other competitors, the coaches, and Shane Adams take on great importance in validating individuals; for instance, Shane saying "Rope is becoming a jousting animal. His aggression ... although he is a soft-hearted cowboy ... will get it done"[55] can authorize a particular sort of masculinity, while Ripper's comments about Paul Suda, "I've had my ups and downs with Paul. He can be a little bit of a pretty boy prima donna, but he can also be a damn good jouster," validates Paul's prowess while calling his masculinity into question, even though he won his joust against Nathan Klassen (who embodied a traditionally American kind of Marlboro Man with a silent cowboy masculinity). Early comments from James Fairclough about Matt Hiltman seeming "a little bit soft compared to the other guys" then get countermanded both by James's triumphing over much larger opponents and Matt's show of prowess in the joust, which led to Shane's saying "my expectations for Matt at this point are quite high."[56] Matt's toughness and talents thus led him to the final match, where he lost by one point to the significantly larger Josh Knowles. While these remarks certainly show the importance of the endorsement of other men, they also demonstrate the show's one semi-radical view of masculinity: size doesn't matter. Despite his endlessly-dwelt-on small stature, James Fairclough claimed his masculine authority immediately by knocking Mike Edwards off his horse in an early practice and continued to be a dominant force throughout, demonstrating the combination of "skill and strength" and "certain aggressive attitude" Karras shows to be necessary for knightly masculinity.[57]

"Knighthood epitomized one set of medieval ideals about masculinity," Karras notes. "The knight competed with other men through physical aggression.... Violence was the fundamental measure of a man because it was a way of exerting dominance over men of one's own social stratum as well as over women and other social inferiors."[58] In *Full Metal Jousting*, this was certainly the case; while Rope Meyers might willingly show his humility, and Josh Knowles might say, after his defeat of Matt Hiltman, "That moment overtook me emotionally, and I seem like a big, tough, guy, but I do have a heart inside this steel carapace,"[59] the show's bread and butter was its valorization of a kind of channeled masculine aggression, a controlled violence that drew both its form and meaning from medieval antecedents finding expression in the contemporary world. As the final episode drew to a close, Shane Adams's final statement cemented this: "It has been 500 years since jousting was in its heyday. These 16 warriors showed that they have the heart and toughness to battle through a brutal competition and now jousting is back."[60]

Thus *Full Metal Jousting*'s medievalism is twofold; the fantasy of chivalry competes with a medievalism in which the Middle Ages stands for an authorized force and abandon. This is the same dichotomy found within chivalry itself as "we must not forget that knighthood was nourished on aggressive impulses, that it existed to use its shining armour and sharp-edged weaponry in acts of show and bloody violence."[61] While outside the lists, humility and generosity may be valued, within the structure of the tournament, the most dangerous, the most violent, and the most aggressive behavior is also the most rewarded. What differentiates this aggression from, for example, American football may be clear from George Will's famous quotation about the latter sport, "Football combines the two worst things about America: violence, punctuated by committee meetings."[62] Jousting becomes popular because it takes away the rules, the committee meetings, leaving only the violence behind. Jousting comes to stand, not for its role in medieval culture as a substitute for battle, but for the battle itself, an opportunity to release oneself from all the confines of contemporary life. The key words here are "stands for"; jousting is not battle. Despite the horrific injuries, no one is truly risking his life, and despite the military rhetoric, no cause, apart from masculine prowess, is being contested. However, at its heart is a desire to be more than mere performance, to achieve a kind of realness, perhaps a vision of Giorgio Agamben's bare life,[63] a life lived—if only in the lists—through pure possibility and pure power, unmitigated by contemporary society's controls. Although it is surrounded by a code of chivalry that confers value on certain kinds of behavioral restraint, at its heart, *Full Metal Jousting* provides a masculine fantasy of the past that revels in violence, drawing on medieval elements (armor, fighting) but assembling them together in ways that serve purposes beyond a desire for authenticity. Shane Andrews and his crew re-create the past not as a real Middle Ages, but as a reflection of their own desires for this past.

Notes

1. Dashka Slater, "Can a Band of American Knights Turn 'Full Contact' Jousting into the Next Action Sport?" *New York Times Magazine* (11 July 2010): 24–29, at 25.
2. *Knights of Mayhem,* 6 episodes, 2011, National Geographic Channel.
3. *Full Metal Jousting,* 10 episodes, 2012, History Channel/A&E Networks.
4. *Ibid.*, episode 1.
5. Juliet Barker notes that "the fifteenth century treatise, Knyghthode and Bataile, which is a paraphrase of the classical De Re Military by Vegetius, says that young men should first be taught to fight by means of the quintain." Juliet R. V. Barker, *The Tournament in England 1100–1400* (Woodbridge, Suffolk: Boydell Press, 1986), p. 150.
6. *Full Metal Jousting,* episode 1, "The Ultimate Extreme Sport," History Channel/A&E Networks, original air date 12 February 2012.
7. *Ibid.*
8. *Ibid.*, episode 6, "Ready to Rock," History Channel/A&E Networks, original air date 18 March 2012.
9. Scott Maslow, "Can Full Metal Jousting Become the Next Ultimate Fighting Championship?" *The Atlantic,* 13 February 2012, http://www.theatlantic.com/entertainment/archive/2012/02/can-full-metal-jousting-become-the-next-ultimate-fighting-championship/252990/, accessed 1 July 2013.
10. Richard Kaeuper, *Chivalry and Violence in Medieval Europe,* (Oxford: Oxford University Press, 1999), p. 22.
11. David Crouch, *Tournament* (London: Hambledon and London, 2005), p. 124.
12. Chrétien de Troyes, *Erec and Enide,* trans. Burton Raffel (New Haven: Yale University Press, 1997), ll. 2112–2118, p. 87.
13. The Battle of the Nations ("Historical Medieval Battle Full Armored Combat"), an annual contest which resembles the tournament *mêlée* much more strongly, is gaining popularity. National teams compete against each other in a variety of competitions (5×5; 21×21) culminating in the large-scale "all against all" battle. Russia has won all four years of the tournament, although the United States team is moving up through the ranks with a fourth-place finish in 2013. For more information see Mike Tierney, "The Holy Grail of Battle Reenactments," *New York Times,* 8 May 2013, http://www.nytimes.com/2013/05/09/sports/battle-of-the-nations-a-holy-grail-of-battle-re-enactments.html?_r=0, accessed 1 July 2013.
14. *Full Metal Jousting,* episode 1.
15. Maslow, "Can Full Metal Jousting Become the Next Ultimate Fighting Championship?"
16. Steven Jeffries, *Tourney and Joust* (London: Wayland, 1973), p. 24. That these Round Table jousts were popularized by Ulrich von Lichtenstein, from whom William Thatcher, the main character in Brian Helgeland film *A Knight's Tale* takes his name, is a particularly interesting detail.
17. Kaeuper, *Chivalry and Violence,* p. 1.
18. *Full Metal Jousting,* episode 5, "Hits Like a Truck," History Channel/A&E Networks, original air date 11 March 2012.

19. Richard W. Kaeuper and Elspeth Kennedy, *The Book of Chivalry of Geoffroi de Charny*, trans. Elspeth Kennedy (Philadelphia: University of Pennsylvania Press, 1996), p. 133.
20. *Ramon Lull's Book of Knighthood and Chivalry and the Anonymous Ordene de Chevalerie*, trans. William Caxton, rendered into modern English by Brian R. Price (Woodbridge, Suffolk: Boydell and Brewer, 2001), p. 16.
21. *Full Metal Jousting,* episode 5.
22. Linda C. McCabe, "Full Metal Jousting: A Review." *Legends of Medieval France and Italy*. 13 (March 2012), http://lcmccabe.blogspot.com/2012/03/full-metal-jousting-review.html, accessed 1 July 2013.
23. *Full Metal Jousting*, episode 1.
24. *Ibid.,* episode 3, "Death Sticks and a Coffin," History Channel/A&E Networks, original air date 26 February 2012.
25. *Full Metal Jousting,* episode 9, "Charge On," History Channel/A&E Networks, original air date 8 April 2012.
26. *Full Metal Jousting*, episode 10, "The Championship Joust," History Channel/A&E Networks, original air date 15 April 2012.
27. Kaeuper, p. 132.
28. *Full Metal Jousting*, episode 3.
29. *Ibid*., episode 5.
30. *Ibid*., episode 6, "Ready to Rock," History Channel/A&E Networks, original air date 18 March 2012.
31. Kaeuper and Kennedy, p. 129.
32. *Full Metal Jousting,* episode 3.
33. *Ibid*., episode 5.
34. Kaeuper and Kennedy, p. 129.
35. *Full Metal Jousting,* Episode 8, "Go to War," History Channel/A&E Networks, original air date 1 April 2012.
36. Kaeuper and Kennedy, p. 129.
37. *Full Metal Jousting*, episode 10.
38. *Ibid.*
39. Rope Meyers, "Crosstraining Journals." http://crosstrainingjournals.blogspot.com. "Courage," 20 March 2012, http://crosstrainingjournals.blogspot.com/2012_03_01_archive.html; "Humility," 12 April 2012: http://crosstrainingjournals.blogspot.com/2012/04/humility.html; "Winning Ugly," 17 April 2012: http://crosstrainingjournals.blogspot.com/2012/04/winning-ugly.html, accessed 1 July 2013.
40. Kaeuper, p. 47.
41. *Ibid*., p. 129.
42. *Full Metal Jousting*, opening sequence.
43. *Ibid*., episode 6.
44. *Ibid.*
45. Slater, pp. 28–29.
46. Kaeuper and Kennedy, p. 123.
47. Kaeuper, p. 153.
48. Slater, p. 29.
49. Clare Lees, "Introduction," *Medieval Masculinities: Regarding Men in the Middle Ages*, ed. Clare Lees (Minneapolis: University of Minnesota Press, 1994), xv-xxv, at xxii.
50. Noel Fallows, *Jousting in Medieval and Renaissance Iberia*. (Woodbridge, Suffolk: Boydell and Brewer, 2010), p. 8.
51. *Full Metal Jousting*, episode 3.
52. *Ibid*., episode 2, "Unhorsed," History Channel/A&E Networks, original air date 19 February 2012.
53. Kaeuper, p. 8.
54. Pierre Bourdieu, *Masculine Domination*, trans. Richard Nice (Stanford: Stanford University Press, 2001), p. 52. Quoted in Ruth Mazo Karras, *From Boys to Men: Formations of Masculinity in Late Medieval Europe* (Philadelphia: University of Pennsylvania Press, 2003), p. 21.
55. *Full Metal Jousting,* episode 2.

56. *Ibid.*, episode 4, "Blood and Guts," History Channel/A&E Networks, original air date 4 March 2012.
57. Karras, p. 37.
58. *Ibid.*, p. 21.
59. *Full Metal Jousting,* episode 10.
60. *Ibid.*, episode 10.
61. Kaeuper, p. 1.
62. George Will, *International Herald Tribune,* 7 May 1990.
63. Giorgio Agamben, *Homo Sacer: Sovereign Power and Bare Life,* trans. Daniel Heller-Roazen, (Stanford: Stanford University Press, 1998).

Bibliography

Agamben, Giorgio. *Homo Sacer: Sovereign Power and Bare Life.* Trans. Daniel Heller-Roazen. Stanford: Stanford University Press, 1998.
Barker, Juliet R. V. *The Tournament in England 1100–1400.* Woodbridge, Suffolk: Boydell Press, 1986.
Bourdieu, Pierre. *Masculine Domination.* Trans. Richard Nice. Stanford: Stanford University Press, 2001.
Chrétien de Troyes. *Erec and Enide.* Trans. Burton Raffel. New Haven: Yale University Press, 1997.
Crouch, David. *Tournament.* London: Hambledon and London, 2005.
Fallows, Noel. *Jousting in Medieval and Renaissance Iberia.* Woodbridge, Suffolk: Boydell and Brewer, 2010.
Full Metal Jousting. 10 episodes. 2012. History Channel/A&E Networks.
Jeffries, Steven. *Tourney and Joust.* London: Wayland, 1973.
Kaeuper, Richard W. *Chivalry and Violence in Medieval Europe.* Oxford: Oxford University Press, 1999.
Kaeuper, Richard W., and Elspeth Kennedy. *The Book of Chivalry of Geoffroi de Charny.* Trans. Elspeth Kennedy. Philadelphia: University of Pennsylvania Press, 1996.
Karras, Ruth Mazo. *From Boys to Men: Formations of Masculinity in Late Medieval Europe.* Philadelphia: University of Pennsylvania Press, 2003.
Knights of Mayhem. 6 episodes. 2011. National Geographic Channel.
Lees, Clare. *Medieval Masculinities: Regarding Men in the Middle Ages*, ed. Clare Lees. Minneapolis: University of Minnesota Press, 1994.
Maslow, Scott. "Can Full Metal Jousting Become the Next Ultimate Fighting Championship?" *The Atlantic.* 13 February 2012.
McCabe, Linda C. "Full Metal Jousting: A Review." *Legends of Medieval France and Italy.* March, 2012.
Meyers, Rope. "Crosstraining Journals." Available online. March 2012.
Ramon Lull's Book of Knighthood and Chivalry and the Anonymous Ordene de Chevalerie. Trans. William Caxton, rendered into modern English by Brian R. Price. Woodbridge, Suffolk: Boydell and Brewer, 2001.
Slater, Dashka. "Can a Band of American Knights Turn 'Full Contact' Jousting into the Next Action Sport?" *New York Times Magazine* 11 July 2010: 24–29.
Tierney, Mike. "The Holy Grail of Battle Reenactments." *New York Times*, 8 May 2013.
Will, George. *International Herald Tribune*, 7 May 1990.

Joan of Arcadia

A Modern Maiden on Trial

STEPHANIE L. COKER

A modern adaptation of Jeanne d'Arc's story, the series *Joan of Arcadia* asserts that "God talks to everyone all the time in different ways."[1] This television series aired during prime time on Friday evenings for two seasons (totaling 45 episodes) on the CBS network from September 2003 to April 2005. The series as a whole and episode 9 ("St. Joan") in particular raise significant questions about the Maid's reception now and in her own day—touching on issues such as her sanity, divine calling, and even corrupt trial. This reading examines the intersection of the modern character Joan Girardi and the historical Jeanne d'Arc to address the lasting impact of the French saint upon a modern audience.[2]

The writers for the show skillfully incorporate many references to Jeanne and the key players in her story. While other television series—such as *Merlin* and *Reign*—attempt to keep the medieval characters in their original time period, *Joan of Arcadia* is exceptional as an homage to the Maid that uses a modern context. Joan, a high-school student with a part-time job at a bookstore, deals with her own family problems as well as those that come with hearing from God. The medieval maiden not only serves as the inspiration for the show's premise, but she is treated directly in episode 9 ("St. Joan") when the protagonist Joan Girardi first learns about the saint in history class. Much to the surprise of her history teacher, the mediocre student Joan Girardi scores an A-plus on the exam and, in a parallel to the actual Jeanne, is then "put on trial" for cheating. Through *Joan of Arcadia*, Jeanne d'Arc finds her place in American pop culture by playing a role in civic education in a television series aimed at teenagers. This essay provides a close reading of the episode "St. Joan" in order to reveal Joan Girardi's dilemma in responding to both her divine visitations and the school authorities—ultimately, the issue of reconciling two realms: faith and education.

Written by Randy Anderson, the episode "St. Joan" offers more than just surface-level similarities with Jeanne's life. The show opens in Joan's history class where the professor is discussing the "tragedy at Agincourt" and how the French soldiers were bogged down in the mud. This scene becomes a metaphor for the entire episode as the characters get stuck in various situations: Joan (in the cheating scandal) and her friends (in the subsequent crusade for her innocence), Joan's father Will (in his job as police chief), and even Joan's professor (who has grown tired of teaching). In my analysis of this episode, I show how the characters, namely Joan, Professor Dreisbach, God, Joan's boss Sammy, and Joan's best friend Grace, all refer—some directly and others symbolically—to historical figures and actual events concerning the famous French maiden. Therefore, I argue that the characters and incidents presented in this episode accomplish what are perhaps the greater goals of series creator Barbara Hall: to educate the audience about Jeanne d'Arc, an exceptional young woman who lived 600 years ago, as well as to emphasize the importance of education.

Before going further, let us consider the dual meanings inherent in the French term *éducation* as it pertains to Jeanne d'Arc's linguistic and historical contexts in medieval France. In her essay "Liturgy as Education in the Middle Ages," Evelyn Vitz points out the twofold sense of the word. In addition to the traditional understanding of *éducation* as school and book learning, the author offers us the etymological sense of the term, "one that is still common in French and several other languages: *upbringing*, the manner in which persons are raised and their character formed."[3] This sense—*upbringing* or *training*—reveals that education depends heavily upon the family and would include faith and specifically—in the case of medieval France—Catholicism. From what is known about the historical Jeanne d'Arc, her life exemplifies *this* definition of education as she was illiterate while exceptionally spiritual from a young age. Centered around faith and devotion to God, Jeanne's education had a lasting impact on her life, as is clear from the historical evidence:

> What do we know of Joan's experience of her religion, aside from the voices she heard? This peasant girl was illiterate. She owned and read no books. She knew by heart a small number of prayers ... all of which are public (liturgical) prayers that were (and are) part of private use. As to the form her religious experience took, this is quite simple: she attended the liturgy with great frequency.... She frequently attended Mass, and during her travels with the army she often went off to nearby chapels to pray. She repeatedly urged her men to pray, to go to Mass, and to go to confession. She would not fight on holy days.[4]

In contrast to Jeanne's sole concern with the spiritual realm, the character Joan Girardi experiences both aspects of *éducation* in the series *Joan of Arcadia*: instruction at a public high school *and* her divine calling and personal faith. Therein lies the struggle for Joan in the modern context as she must reconcile faith and reason. She clashes, for example, in the "St. Joan" episode with her history teacher whose spiritual skepticism is apparent in his version of Jeanne's story, discussed below.

Trying to reconcile faith and reason and questioning whether or not the two can coexist is not a new topic of discussion. For centuries, philosophers have debated this topic, and I mention here only two to demonstrate the conviction found on both sides of the issue. During the Enlightenment, Voltaire asserted the impossibility of faith coexisting with reason. In his *Dictionnaire philosophique* (1764) he defines faith in the following manner: "[Faith] can be nothing but the annihilation of reason, a silence of adoration at the contemplation of things absolutely incomprehensible" ["Elle [la foi] ne peut donc être qu'un anéantissement de la raison, un silence d'adoration devant des choses incompréhensibles"].[5] Thus, according to Voltaire, if man embraces faith, he forfeits his ability to reason. Less than a century before Voltaire, French mathematician and philosopher Blaise Pascal addressed the topic of faith and reason in *Pensées* (1670). For Pascal, the two concepts not only coexist, but both are necessary and work in tandem. Concerning faith, Pascal states that the heart—not reason—experiences God ["C'est le cœur qui sent Dieu et non la raison. Voilà ce que c'est que la foi: Dieu sensible au cœur, non à la raison"]. He also exposes the limits of reason by noting that the final step in reasoning is to recognize that there is an infinity of things that surpass it ["La dernière démarche de la raison est de reconnaître qu'il y a une infinité de choses qui la surpassent; elle n'est que faible, si elle ne va jusqu'à connaître cela"]. Pascal, however, acknowledges the merit and the need for reason to complement faith when he declares that there are two excesses or faults: excluding reason and accepting *only* reason ["Deux excès: exclure la raison, n'admettre que la raison"]. The perspectives of French philosophers from the seventeenth and eighteenth centuries beg the question as to what would have been the consensus about faith and reason/education in Jeanne's day.

While Voltaire states that faith annihilates reason, medieval philosophy asserts instead that *reason annihilates faith*. In order to be devout, one ought to eschew formal education as it could corrupt Christian values. Evelyn Vitz addresses this notion inherent in medieval education when she writes:

> Book learning has never been thought of as necessary to salvation or, indeed, to sanctity. The Catholic and Orthodox traditions contain many saints who could not read or write; *some of these figures positively, actively, refrained from acquiring such learning*.... In Protestantism, literacy and Bible reading have been, from the very

> start, important for all believers.... The Catholic and Orthodox traditions have had mixed feelings about the importance and value of book learning; *they have been concerned by the dangers presented by the world of the school and students and indeed of learning itself.* This is not, of course, to deny that there have been great scholars among Catholic and Orthodox saints and respected figures, but ambivalence toward the value of learning is unquestionably present.[6]

The perceived threat and, therefore, imposed distance between the two realms of faith and learning that was manifest in Jeanne's era still exists today and, as previously mentioned, it becomes a focal point for the series *Joan of Arcadia.*

Series creator Barbara Hall's own experience may explain why the show concentrates on this topic. In a 2005 interview, she revealed how her personal journey to faith (and Catholicism) came through, among other things, the study of religion and science: "I had a kind of near-death experience that gave me an understanding of something bigger than myself. This was a journey of discovery during which I studied every world religion. It didn't take the form of coming to the Catholic Church or any particular religion for a really long time."[7] When asked why "*Joan of Arcadia* reflects a keen awareness of science," Hall responded that science greatly influenced her decision to pursue faith:

> During the time I was searching, I developed this inexplicable fascination with physics. I read everything I could get my hands on about physics and theology. At the end of that search, I decided I wanted to go back to the life of faith.... So, *in terms of my conversion, I credit science more than anything with my becoming a Catholic.* My favorite book on physics right now is one by Stephen M. Barr, *Modern Physics and Ancient Faith*.... He explains how ludicrous it is to think that science and God are enemies.[8]

Thus, it is not surprising that Hall's characters confront crises of faith in the course of the series. They question God when tragedies occur—such as when Joan's older brother Kevin, a star athlete, becomes a paraplegic after a car accident. Throughout the series, Joan's family members discuss matters of faith, and they even express concern when Joan becomes obsessed with Jeanne d'Arc. In this instance, the family discussion at dinner turns to saints because Joan is reading *Butler's Lives of the Saints* at the table. Her father states how "the Church has to have a saint for everything" (1.9).[9] In response to his sarcasm, Joan's mom Helen explains that their purpose is to "show us how to live," but the youngest, Luke, is not satisfied. He comments that great scientists have been denied—or not considered for—sainthood, and his remark underscores the theme of separation between faith and reason: "I find it unconscionable that science martyrs never get the proper respect. Like Galileo said the earth revolves around the sun and he was put under house arrest, but is there a Saint Galileo?" (1.9). Luke's case for Galileo continues when he argues that the scientist "was brought up on charges and forced to recant" (1.9). This fact about Galileo's life bears a striking resemblance to the story of Jeanne who recanted during her trial in 1431 (before ultimately retracting that statement and being sentenced to death). It is necessary at this juncture to rehearse some historical details pertaining to Jeanne's life in the 15th century in order to highlight the parallels that emerge in this episode of *Joan of Arcadia*.

Circa 1412, Jeanne d'Arc was born in Domrémy, a small village in northeastern France. As a young girl (probably at the age of twelve), Jeanne had her first angelic visitations. The trial records state that the voices that guided the Maid were those of Saints Michael, Margaret, and Catherine. In the beginning, they instructed Jeanne to attend church and pray. As she grew older, the Maid received a specific mandate to visit the dauphin Charles (son of Charles VI). In February 1429, at the age of seventeen, Jeanne traveled to Vaucouleurs to visit the Duke of Lorraine before journeying to Chinon to seek an audience with Charles. Reluctant to believe the Maid, the dauphin had her examined by theologians at Poitiers. After three weeks of interrogations and a physical examination, Jeanne won the council's approval. The dauphin Charles then appointed her *chef de guerre*, and Jeanne proceeded to Orléans in late April 1429. Jeanne's lifting of the siege at Orléans began her significant, though short-lived (only two years), military career. The Maid's victory in May 1429 also marked a turning point for France in the Hundred Years War, and this event led immediately to Charles VII's coronation. Crowned at Reims in July 1429 with the Maid at his side, Charles VII was officially recognized as the new French monarch. Jeanne, captured by the English in 1430, was put on trial and accused of heresy. In May 1431, Jeanne was burned at the stake in Rouen, France. In 1456, she was retried posthumously and declared innocent in the rehabilitation, or nullification, trial. Not until 1920 would Jeanne d'Arc be officially recognized by the Catholic Church as a saint.

Since the 15th century, Jeanne d'Arc has inspired countless authors from Christine de Pizan to Voltaire, and she holds a special place in the collective French memory. In his multi-volume work *Realms of Memory: Rethinking the French Past,* Pierre Nora treats the traditions, ideals, and places that constitute the notion of France and French-ness. In the section entitled *Identifications*, the editor includes Jeanne d'Arc as one of three great symbolic figures in French history (and the only female). As Michel Winock notes, the iconography of Jeanne persists in all arenas:

> [The] name Joan of Arc has lent itself to a variety of purposes since the nineteenth century. It has been used to sell mineral water as well as to distinguish Catholic youth groups and political organizations. The Maid has been portrayed in a variety of stereotypical attitudes: listening to voices, delivering Orléans, dying in Rouen. Her image has been associated with official ceremonies, local festivals, post cards, church windows, almanacs, picture books, and souvenirs for pilgrims and tourists....

> Thousands of statues depicting her as an inspired shepherdess or intrepid warrior can be found in every village in France, decorating monuments to the dead, tucked away in church ambulatories, or lording it over town squares.[10]

The Maid's image pervades France: from the smallest villages to the capital city of Paris, memorials to Jeanne are ubiquitous. Her story has also been featured numerous times on the silver screen. From one of the first filmmakers, Georges Méliès, to many other notable *cinéastes* including Cecil B. De Mille, Carl Theodor Dreyer, Victor Fleming, and Robert Bresson, Jeanne's life remains a popular choice among directors.[11] In "Joan of Arc and the Cinema," Robin Blaetz notes that few of these films rely on what she calls "*mysteries of faith in a non-religious age*. The majority of the films about Joan, and certainly all those made in Hollywood, *are driven instead by the appeal of the spectacle inherent in the story* and the paradox of Joan of Arc in relation to women."[12] Even on the small screen in both the U.S. and France, Jeanne became the subject of several television movies, and the Canadian television mini-series *Joan of Arc* (1999), starring Leelee Sobieski, was broadcast internationally and recognized by critics, earning four Golden Globe nominations. The American series *Joan of Arcadia*, however, stands alone as a television series that focuses on a modern version of the Maid. When it premiered, the hour-long drama was well-received by both public and critics. The American Film Institute named *Joan of Arcadia* among the top ten television programs of 2003; it was also nominated for the 2004 Emmy awards "Outstanding Drama Series" and "Outstanding Lead Actress in a Drama Series." In its first season, the show averaged over ten million viewers, and *Joan of Arcadia* won the People's Choice Award for "Favorite New TV Dramatic Series." But if *Joan of Arcadia* offers the audience no sign of medieval armor, no dramatic combat scenes on the battlefield, and no tragic death at the stake, wherein lies the intrigue that is expected when it comes to television medievalisms? The appeal of *Joan of Arcadia* is found in its refusal of the spectacle in Jeanne's story in favor of confronting the "mysteries of faith in a non-religious age" by focusing on notions of theology, divine calling, the voice of God, and the reception of His voice.

Unlikely Allies: Joan Girardi and Professor Dreisbach

While the lead character Joan Girardi is the middle child in the Girardi family (who just recently moved to the fictional town of Arcadia, Maryland where her father is the new chief of police), she is not the average American teenager. This 16-year-old converses with God who appears in various places—at school, on the bus, and even on her TV. God has an assignment for Joan in every episode, and the God-character takes on many different human forms including a little girl, a cute teenage boy, and a lunch-lady in the school cafeteria.

The episode "St. Joan" begins in Joan's history class with a lecture on the Hundred Years War. The scene opens with students viewing the teacher's slideshow and listening to his commentary: "The English longbows rain arrows on the French knights in their armor. Angry, the French charge across the ploughed field. Then, tragedy. What tragedy? Class, who knows?" (1.9). Here, Professor Dreisbach introduces students to the significant loss the French suffered at Agincourt. During his lecture, the teacher catches Joan Girardi sleeping, but star student Steven Zakheim responds, "The French soldiers and horses floundered in the quagmire. They got slaughtered by the Brits." At this point, let us examine this pivotal moment at Agincourt in order better to understand the two factions that existed in France at that time and to underscore how the professor's own classroom reflects this division. Describing the dismal state of France, Jeff O'Leary explains the devastating loss at Agincourt:

> The French had fallen from the heights of national prestige since the loss of thousands of their best knights facing King Henry V at Agincourt in 1415. The French defeat left the Dukes of Burgundy, Normandy, and Brittany in the north to fend for themselves, and various pacts were arranged to keep their power and lands against a potential English invasion. These pacts protected the dukes but depleted any hope for a united France.[13]

Of the various political divisions, two predominant factions emerged: the Armagnacs and the Burgundians. As the Burgundians allied with the English, the question of who would succeed Charles VI as king of France became a complicated one.

Initially, it seemed that Charles VI's son would become the next French king; but Charles VII encountered many obstacles along the way and his kingship began to look unlikely. In 1418, the Burgundians entered Paris and massacred the Armagnacs. Among those who fled the city was the dauphin Charles. Then came the Treaty of Troyes and, with it, the end of Charles VII's royal aspirations. With this treaty, Charles VI disinherited the dauphin, and Henry V of England was recognized as next in line for the French throne. In 1422, Henry's wife Catherine gave birth to a son, Henry VI. When both Henry V and mad Charles VI died in 1422, little Henry VI was proclaimed King of France and England. At this moment however, the Armagnacs—who were resistant to English rule—stood behind the disinherited Charles. They recognized him as the rightful heir to the French throne and submitted to his rule as "Roi de Bourges" in the south of France where Charles had fled after the Burgundian invasion of Paris. Whereas England and Burgundy wanted Henry VI of England to be the next French king, the Armagnacs felt that Charles VII should assume the French throne. At the end of the Hundred Years War, France needed to liberate itself to create a national identity. A newfound hope was necessary to rectify the tense political division between the Armagnacs and the Burgundians. This hope would come in the form of a maiden, Jeanne d'Arc.

The *Pucelle*—as she called herself—was not, however, unanimously accepted by the public. Divided in their opinions of Charles VII, the people of medieval France similarly struggled with Jeanne's military role. This particular moment of ambivalence and division is captured in Dreisbach's class lecture in the *Joan of Arcadia* episode. The professor's discussion of the Hundred Years War turns to the legendary figure of Jeanne d'Arc, and he clearly expresses his own skepticism of Jeanne. While the professor credits the Maid with a significant victory, he also condemns her to madness:

> After the debacle of Agincourt, the French [were] humiliated, divided, conquered, and then, to save the day, comes.... Jeanne d'Arc. Or as we know her, Joan of Arc, a peasant girl who talks to God or so the legend says. Now, talking to God was not unheard of back then, but, uh, Sigmund Freud would've provided Joan of Arc's parents with a different analysis: paranoid schizophrenic with a messianic complex [1.9].

His Freudian reading of the Maid captivates Joan Girardi, and she interrupts her teacher to exclaim, "Wait! She was crazy?" (1.9). To which the professor replies: "God told her to get together an army to save France from the British. I think we can draw the necessary conclusions" (1.9). Dreisbach clearly doubts Jeanne's divine calling, and as he conveys the story to the class, Joan becomes frustrated and challenges the professor's interpretation, "That doesn't mean she was crazy just because she talked to God." Here, the protagonist Joan feels the need to defend the French maiden whose story—filled with visitations from God—sounds much like her own. After studying, she confronts her professor again in history class; and, moments before the test begins, she accuses him of not recounting the Maid's true story:

> I did study, and I learned stuff. Like, Joan of Arc wasn't a schizo, for example. And they didn't kill her for saying she heard voices. They killed her for wearing pants.... Her trial was totally corrupt. She was a scapegoat.... *You're teaching the whole thing wrong.* (pause) *Well, what really happened is*, like, these bossy judges forced her to wear pants in the courtroom. Made it look like she was a witch, which totally gave them permission to fry her [1.9, emphasis added].

Joan's outburst can be read as yet another defense of the Maid, as she directs this announcement to the class and also to her professor whom she believes is "teaching the whole thing wrong." Joan's accusation against Professor Dreisbach and her subsequent explanation of "what really happened" renders his character as a parallel to the dauphin Charles. He knows about the Maid, but doubts her credibility. Joan Girardi tries to convince the professor of the Maid's true story—much like the Maid herself must convince the dauphin of her own divine mission.

When Joan Girardi announces that Dreisbach is "teaching it wrong," she is, in fact, challenging his authority and implicating the professor whose misinformation in class perpetuates false opinions about the Maid. Here again, the professor reminds us of the dauphin Charles—trying to regain the French crown—as he too has lost some of his authority. Dreisbach even acknowledges this fact in his response to Joan: "Apparently my authority counts for nothing. All right. You may open your tests and begin" (1.9). Immediately following this statement, the class—including Joan—begin testing. To draw yet another parallel to the Maid's story, this moment of examination could be read as an allusion to Poitiers where Jeanne had to prove her authenticity.

The most compelling evidence for the comparison between Dreisbach and the dauphin Charles occurs at the end of the episode when Joan discovers that her professor is bogged down in his despair over teaching. Dreisbach's discourse reveals the dramatic impact Joan has had on him after succeeding on the exam (twice); and the professor's references to the real Jeanne are obvious here:

> Somewhere along the line, I got discouraged and I started ... phoning it in. I'm aware. It's a teacher's greatest fear. Before this event, I was going to quit. This was going to be my last year and it was causing me a lot of pain because *I wasn't going out in a blaze of glory. I was surrendering in defeat, like the French at Agincourt, floundering in the mud of my students' indifference.* But I made you care about history, Ms. Girardi. I don't know how I did it, but I did. And that's the whole point. *You inspired me to take back my crown.* I thank you [1.9, emphasis added].

In his speech, the professor compares himself explicitly to the dauphin Charles and thus likens his student Joan to the historical Jeanne. The portrayal of Joan and her teacher as unlikely allies parallels the historical moment when a dauphin needed the help of a young girl to become king.

State of Grace

The connection between Jeanne's and Joan's revolutionary behaviors—especially evident in this episode—lies in their non-conformity to expectations, resistance to authorities, and testing. Historically, Jeanne's testing played an important role in determining her eligibility as a divine agent. Proof of purity ensured Jeanne's authenticity as *la Pucelle* and thus demonstrated that she was an untainted, pure vessel. In the case of Joan Girardi, her testing yields doubtful results: she aces the history test but, rather than confirming her credibility, Joan's examination raises questions for Professor Dreisbach who assumes that the mediocre student has cheated on the exam. During a conference, Dreisbach and the principal confront Joan and ask her to explain the anomaly of her recent success on the test: "Joan, the highest score you ever got on a history test was C-minus" (1.9). Considering this trend, they wonder how it is possible that Joan earned an A, and they ask her to retake the test. Initially, Joan expresses concern that her retaking the exam will be misconstrued by everyone as an admission of guilt: "Agreeing to retake the test, aren't I admitting I'm a cheater? I'm not going to admit to that because I'm not!" (1.9). When the news spreads of Joan's stand against the school authorities, her allies begin a protest on the front steps of the school. Led by Joan's best friend Grace, the revolution to save Joan involves wearing buttons, handing out flyers to fellow students, and carrying signs like the one that reads "No proof! No test!" Grace encourages other students to join the cause as she shouts to passersby: "This affects you, people. You could be next! Rally against injustice! No proof, no test! Support Joan Girardi against the neo-fascists. Rally against injustice! Support Joan! Take back the power! Take back the power!" (1.9). When Joan arrives and sees the demonstration Grace has organized on her behalf, she fears the principal's reaction. Grace reassures Joan and offers her some guidance about how to proceed: "It's okay. You've taken a stand. I'll organize the effort.... The slogan needs a little work. I threw it together sort of.... So did you put together a list of your demands? You have to have a list of demands. It's, like, in the revolutionary handbook" (1.9). Although hesitant at first, Joan is soon convinced by Grace to continue the revolution: "Do you believe in the writ of habeas corpus? ... They have to have evidence to accuse someone of a crime.... It's the foundation of a free society.... You took the stand. That's the hard part. Let's take this baby all the way. Are you ready, girl warrior?" (1.9).

Here, when Grace explains legal matters such as *habeas corpus*, background music enhances the connection to historical events in Jeanne's life by recalling the Maid's final moments at her trial and execution.[14] At this pivotal moment when Joan decides to join in the protest, the audience hears the musical introduction of "Time Has Come Today" as it builds behind the actors' voices. Then, Joan Jett belts out the chorus which underscores Joan's decisive move. The song lyrics insist that the time is now, this moment is inescapable, and that the end may be tragic. The individual in the song even acknowledges that s/he may end up in flames.[15] This song contributes to the overall climactic effect of Joan's call to arms; furthermore, when the lyrics assert the impossibility of escaping, they evoke images of Jeanne's imprisonment before and during her trial in Rouen. Indeed, when the songwriters indicate the likelihood of death by fire, they not only remind the audience of Jeanne d'Arc's destiny—her death at the stake—but the use of the first-person pronoun "I" accentuates the Maid's own realization of her inevitable fate in 1431. Even the song's refrain mentions how someone who was once loved has now been forgotten. These lyrics suggest notions of the Maid's popularity that peaked and waned during her lifetime (and perhaps, still today). But ultimately, for Joan Girardi, it is the repetition of the word *time* that emphasizes the timeliness of her decision to join the revolution. She must act now, join the protest, and, as Grace puts it, "send a message to the despots" (1.9). However, much like Jeanne's military career, Joan's "revolution" is short-lived and after her third visitation from God, she decides to call off the school protest. Following his advice, Joan retakes the test and proves her innocence. Her friends, however—especially Joan's best friend, Grace—are upset that she backed down. Having organized the protest and assembled the participants, Grace becomes frustrated with Joan's decision to end the revolution.

The character of Grace is of particular interest to Karin Beeler whose book, *Seers, Witches, & Psychics on Screen*, examines women visionary characters in recent television and film. In the chapter, "Teen Visions of God: Postfeminist Heroism and Genre Crossing in *Joan of Arcadia*," Beeler notes that Joan's best friend Grace, who is steadfast in her revolt against the authorities, is more like the historical Jeanne d'Arc than the Joan character:

> In the episode "St. Joan" [1.9], Joan actually explores similarities between Joan of Arc and her own life.... Her friend Grace encourages her to engage in a political battle and calls her a "girl warrior" [1.9]. Joan, however, is a rather reluctant leader or warrior figure. She does not have the same spirit of resistance as either Joan of Arc or *her radical friend Grace, who turns Joan's test situation into a school-wide crusade and ends up acting more like a Joan of Arc figure than Joan does*. After God convinces her to revisit her original decision to not rewrite the test, Joan ends up losing Grace's respect and the support of her friend Adam who had also rallied behind Joan.[16]

While I agree with Beeler's observation that Grace displays many Johannic characteristics, I must point out the significance of Joan's determination to fulfill the divine assignments she has been given. When Grace (and others) see Joan Girardi's decision to retake the test as a choice to back down, she sees it as being faithful to her divine mission and following the counsel she received from the three God-characters in this episode. Even when the school principal ridicules Joan for giving in and retaking the test, she exclaims: "I didn't cheat. I really studied. I didn't start this revolution, and I didn't want to back down either. This is all just bigger than me, so please *let me do it the hard way*" (1.9, emphasis added). Here, she echoes what God had explained to her in the third visitation: "martyrs did things the hard way" and, in submitting to testing, she is behaving more like Jeanne than the Grace character.

Both Joan and Grace have an undeniable connection to Jeanne d'Arc. Grace regards Jeanne as superior to all other heroines in history when she states, "Joan of Arc was, like, *the* girl warrior. Strapped on chain mail and led men into battle. Naturally, they burned her at the stake" (1.9). Grace discusses her admiration for the French maiden, and even her name *Grace*, with its spiritual connotation, offers another reminder of Jeanne's life and trial. The concept of grace was clearly at work in Jeanne's life from an early age when she was called to set herself apart and go often to church. The Maid's divine separation and sanctification continued throughout her lifetime, and the notion of grace recurs at an important moment during Jeanne's trial. When asked by her judges if she is in a state of grace, Jeanne answers, "If I am not, may God put me there; if I am, may He keep me there."[17] This response concerning her "state of grace," now a celebrated scene from Jeanne's trial, is often replayed in Johannic films and shows how the Maid confounded her judges.[18] Thus, it is fitting that the series' writers created the character of Grace to be Joan's sidekick and to assist her. The two of them, both outsiders, must find their way amongst their peers. In her book, Beeler comments on the young women's unconventional nature:

> Like the historical Joan of Arc, Joan Girardi is a non-conformist. She hangs out with two eccentric characters, Adam, a solitary artist figure, and Grace, a girl with few social skills who is labeled a lesbian. Just as Joan of Arc was considered unusual and demonized and eventually declared a witch, Joan Girardi is constructed as a freak in the opening episode. However, she is not a freak in isolation. Her "difference" or otherness as a girl who has visions of God is presented in the context of other "freaks" in the series and serves as a metaphor for a teenager's fear of not fitting in.[19]

Both Joan and Grace pride themselves on non-conformity, and they recognize this trait in their idol, Jeanne d'Arc. When discussing the French maiden's alleged insanity, Grace explains that the problem lay in her unwillingness to adhere to the gender-standards of her time: "Anytime you deviate from the norm, the fascists call you crazy. I consider it a badge of honor" (1.9).[20]

While both female characters emulate Jeanne, it is ultimately Joan Girardi who most closely resembles the medieval heroine when she saves Professor Dreisbach. Beeler credits Joan with a *private*—rather than public—act of heroism when she inspires her teacher:

> [Joan] does symbolically save the life of her teacher, Mr. Dreisbach, who had almost given up on teaching until Joan's success on her second exam convinced him to rethink his earlier decision. The episode reveals how being a public heroine can be less important than saving "someone's life" [1.9] in a more private manner. This private act can also be a form of resistance. In other words, the life of an individual can displace the symbolic value of a political cause...[21]

According to Beeler's reading, a new hierarchy is established: the private heroic act of saving an individual supersedes the greater political cause. For Beeler, Joan's revolution at school becomes less important than saving her teacher. I argue, however, that Joan's impact on Dreisbach produces change that affects more than just one individual. Hers is a private act of heroism that has broader implications: this small, personal victory affects a greater cause once we consider not only the professor's sense of fulfillment but also that of his students. Likewise, the Maid's triumph was not just an individual victory for Charles, but one with far-reaching results for France.[22]

Éducation : What's Old Is New Again

The episode "St. Joan" not only offers insight into contemporary attitudes towards the medieval maiden but also comments on the current educational system and the importance of learning. The characters reveal current judgments—namely, skepticism—of Jeanne d'Arc's sanity and her divine calling while still insisting that the audience explore history books to learn more. Joan's first divine visitation, for example, occurs when Gardener God tells her that ignoring history brings consequences: "Here's what I want you to do.... This history test that's coming up, I want you to ace it.... History is important.... Last spring, they didn't cut the branches back. This summer, they had too much shade. Now they've got too many leaves. The grass is dying. They have to replant the lawn. That's what happens when you ignore history" (1.9). Joan objects by offering the all-too-familiar excuses: "But I hate history, and this is the Hundred Years War, which was really long.... Ohhh, but Dreisbach is so boring. And history is, like, so over!" (1.9). To challenge this disinterest in history, the writers of *Joan of Arcadia* ingeniously create a storyline which discusses and alludes to the actual events of 15th-century France, and they utilize a cast of characters including Joan, Grace, God, and Dreisbach to reenact Jeanne d'Arc's story and show its relevance to a modern audience.

Taking into account that many of their young viewers may be unfamiliar with the story of Jeanne d'Arc, the writers provide plenty of background information. At the bookstore, for example, when Joan Girardi asks her boss Sammy if he will tell her about the famous French maiden, he offers a brief summary and, in doing so, cannot hide his disapproval of the younger generation who are less informed about history:

> I can't tell you about Joan of Arc, because in order to do that, I'd have to assume you have some basic grasp of anything that happened before, say, the Reagan years.... Well, here's an interesting approach. Read the books.... Joan of Arc met with the dauphin, told him he should assume the crown and take back France from the British. Then she got an army together and made that happen. A teenage girl who managed to drive the British out after years of occupation, and you have trouble working your iPod [1.9].

In challenging Joan to read books and stay informed, Sammy's critical remarks seem to target a larger audience, perhaps teenagers in general. The didactic tone of the series as a whole comes across in his advice to his employee: "Joan, I have a Master's degree in English literature. I could've done a number of things with my life. (pause) At least three things, but I chose to open a bookstore, because I believe in the power of knowledge, which comes from books. You want to learn something? Read" (1.9). Sammy attempts to shame Joan into reading books and, like Professor Dreisbach, has certain expectations for what Joan and her peers should already know about the Hundred Years War and the Battle at Agincourt. For these two authority figures, it is not about teaching a younger generation, but rather about criticizing their ignorance. Neither character seems to consider the possibility that they should inculcate an *interest* in learning or actually teach their students from scratch.[23]

Other school authorities—the principal Mr. Chadwick and assistant principal Mr. Price—seem to parallel Sammy and Dreisbach by proclaiming the importance of education while threatening the students for supporting Joan. After Joan retakes the test, the principal confronts the protest participants. He tells Grace that this rebellion must end so that school can resume: "Ms. Polk, I appreciate your effort to launch your lifelong career as a free radical, but I have a school to run, board of directors to answer to, not to mention all of your parents. *I promised their children an education, and I'm going to make sure they get one*" (1.9, emphasis added). Grace responds by challenging the principal's logic, asserting that their revolution *is* an education when she inquires, "Civil disobedience is not an education?" (1.9). Mr. Chadwick, annoyed by Grace's rebellion, answers her question with the snide remark: "We have big thick textbooks that talk all about that, if you'd bother to crack one" (1.9). Here again, the school authorities emphasize reading books (not teaching) to understand history while Joan and her friends prefer asking questions and taking action to pursue justice for the wrongly accused. The decision by Joan and Grace to take on the school authorities harkens back to the notion of faith as a call to action. This view of faith as leading to an active response was typical of the Middle Ages: "Men and women understood their faith ... as calling for something on their part: this action could be imitative, as the life of Christ, the Virgin, or the saints; or it might involve initiatives of a moral, physical, or esthetic nature, understood as done for Christ, the Virgin, or a saint."[24] This is certainly the case for Joan and Grace who believe their revolution imitates what Sainte Jeanne would have done by taking a stand against the authorities.

Their "call to arms" is short-lived, however, when both faculty and administration attempt to quash the revolution. In an effort to dissuade Grace's collaborators, Mr. Price pressures the group of adolescents into rethinking their dedication to Joan's cause and threatens them with suspension from school: "What about the rest of you? *Are you willing to die for your beliefs? Symbolically?* Because anyone who leaves this school wearing one of those buttons will be suspended until further notice" (1.9, emphasis added). Thus, the show *Joan of Arcadia* villainizes the current educational system that relies on threats and power rather than inspiration. Before Mr. Price issues this warning to the students, Joan's close friend Adam backs down and deserts Joan. This scene, like several others discussed here, evokes the historical account of the life of Jeanne d'Arc who—

after being captured by the English—was abandoned by her French allies and the newly-crowned King Charles VII. Even the principal's query "Are you willing to die for your beliefs?" reminds the audience once again of Jeanne's tragic fate.

Earlier, this essay discussed the French term *éducation* in the broad sense of not only book learning, but also as it pertains to upbringing and character formation. Through the characters (namely Dreisbach, Sammy, and the two principals), the notion of educating by books alone continues to fall short. Consequently, Joan and her peers become frustrated by this type of education, as they must also grapple with spiritual matters for which they feel unprepared. The divide between reason and faith comes to the forefront and highlights the need for a type of spiritual education. Incorporating the religious aspects of divine calling and faith, the series promotes educating the whole person—both mind and spirit—and thus, encourages a return to medieval spiritual ideals.

The Theology of Joan of Arcadia

In the *Joan of Arcadia* series, the God-character(s) play a unique role in their interactions with Joan. The deity takes many different forms, and Joan never knows when or in what form to expect Him/Her. Their surprising and ambivalent appearances contrast the predictable, book-bound forms of education represented by school authorities and press Joan to experience and explore the relationships among knowledge and action, offering a fuller and more integrated concept of education. This section examines the three God-characters introduced in episode 9, Joan's reception of them, and how they reflect the overall theology presented in the *Joan of Arcadia* series to promote the notion of whole-person education. In her article, "Can Television Mediate Religious Experience?: The Theology of *Joan of Arcadia*," Angela Zito examines the way in which the show's creator and writers carefully establish the mysterious quality of God:

> God takes the initiative, appearing to Joan unexpectedly, speaking through random people she encounters, and assigning her mysterious tasks, whose reasons only become clear as the plot unfolds. Neither Joan nor the audience knows what will happen. God has a plan; Joan has doubts. God is like the writer; Joan is the actor who (along with the audience) must make sense of the script.[25]

In addition to giving God a human form, Barbara Hall and the writers underscore the omnipresence of God as He appears often and in the most unlikely of places. In the episode "St. Joan," for example, three different God-figures speak to Joan: a Caucasian male groundskeeper at school, a young Hispanic woman knitting on a park bench, and a Caucasian male janitor repainting the school walls. Their messages for Joan will soon be addressed in more detail, but it is important to note the ordinary status of those who act as divine spokespersons.

Throughout the series, the writers do not select a priest or even Joan's parents to be the voice of God. Instead, it is the marginalized individual—often overlooked and unseen—who personifies the deity, and his/her appearance onscreen is devoid of fanfare or the blinding light one might expect. In her article, Zito notes "the show's devotion to the quotidian" and she points out the results of this conscious choice: "its absolute rejection of special effects to carry divinity to the viewer ... presents a picture of God as available to us daily."[26] And, I would add, presents a picture of education as available in surprising, quotidian, action-oriented moments.

Of Joan's three visitations from God,[27] her first dialogue with the divine turns to the historical Jeanne and her sanity. When Gardener God advises Joan to study history, she redirects the conversation to ask him about the Maid: "Let's get back to the crazy thing, ok? Were you really talking to Joan of Arc? And am I—am I like her?" (1.9). But, at this point, God will only tell her that she must succeed on the history test. It is her first mission, and Joan thoroughly prepares herself for the exam. The God-characters, in addition to giving orders to Joan, also reprimand the protagonist. In a scene where Joan waits at the bus stop, she encounters God again—the second visitation in the form of a young Hispanic woman.[28] The message is simple, and God scolds Joan for showing disrespect towards her teacher. She/God tells Joan: "If you make snap judgments about people and are unwilling to look into their past, you'll never begin to understand them.... He's your teacher.... You don't have to like him. Let him teach you" (1.9). Primarily, this message from God reinstates Dreisbach's diminished authority; and secondly, this divine visitation upholds the show's ultimate goal, as stated above: to emphasize (and reframe) the importance of education. Here, God praises Joan's teacher, and also adds value to the subject of history by encouraging Joan and the audience to "look into [the] past" in order to really understand.

Joan's third conversation with God in this episode reveals another aspect of God's nature that is seen throughout the *Joan of Arcadia* series. While Joan typically struggles to understand the reasons why she must complete a certain task, the God-character—his desires and plans—are often inexplicable.[29] This time in the form of a middle-aged janitor, the God-character advises Joan to submit to retesting when school authorities suspect Joan of cheating and explains that "[martyrs] did things the hard way" (1.9). Initially, Joan rejects his advice and tries to persuade him that she is acting just as Jeanne would in this situation: "What do you mean? You said study history, so I did. You said get an "A" on the test, so I did. Along the way, I learned about Joan of Arc and figured out the whole martyr deal. And now my word is being challenged. Like hers. I'm taking a stand. It's perfect" (1.9). Her efforts to reason with God, however, are to no avail, and he responds to Joan by alluding to his mystery and incomprehensibility: "I'm not really here to discuss martyrdom with you, Joan. *Like most things having to do with me, it's complicated.* Retake the test" (1.9, emphasis added). Joan's ability to follow rote rules must be complemented by faith, by a surrendering of reason to this command.

Next, let us consider Joan's reception of the divine messengers—a response that often involves incredulity and sarcasm, highlighting her own commitment to reason over faith. As Joan meets the varied representatives of God in each episode, she initially questions and resists God's orders. Joan's uncertainty usually delays her from carrying out her divine mission because she frequently does not understand the reasoning behind the mandate, and the audience thus sees the character's inner struggle between doubt and faith. The show's creator Barbara Hall exposes this conflict as a natural response and one that is especially typical of a teenager. In an interview, Hall discussed her childhood idol and revealed that her own daughter serves as the inspiration for Joan Girardi:

> When I was a young girl, I looked around at Cinderella and Snow White and all those girl icons and could not find myself anywhere. The only place I did find myself was with the form of Joan, the girl warrior. She continues to fascinate me. When my daughter was about to come of age, I began to wonder what it would look like if God tried to grab the attention of a teenager today. What would that process look like, and would a modern teenager be able to have the fortitude to follow that calling? Then I sort of pictured what my daughter would do if God ever talked to her, and I knew she would be the type to argue with God, getting snippy.[30]

It is this snippy and sarcastic nature that characterizes Joan Girardi, played by the actress Amber Tamblyn. Joan's sarcasm, which dominates her personality, surfaces in all of her daily interactions: with her family, her friends, the school authorities, and her boss. Since the God-character appears to Joan in *human* form, it is quite natural that she also addresses Him in a sarcastic tone, rather than with reverential fear. If He appeared to Joan in angelic form—as the story goes for the historical Jeanne—the lead character would perhaps respond differently and find the messenger and, therefore, the message more credible. Joan, however, is usually surprised and suspicious when God interrupts her daily life with a new directive: "when God appears to Joan, the first indication of presence is usually some stranger calling her name. She is hailed in that most intimate of ways—God always already knows her. As audience we are always already ready for this display of interpellation as she responds to the hailing."[31] In fact, it is the ordinary, human form of the divine that confuses Joan when she tries to discern the voice of God. At one moment in the episode, Joan mistakenly assumes that her family therapist is another manifestation of God. When the therapist mentions that she has assignments for the family, Joan suspects that she is God and asks her for a sign: "Wait! We're going to have assignments? I can't have any more assignments. Are you a therapist or are you.... *You* ? Blink twice if you know what

I mean" (1.9). For Joan's family, the outburst calls her sanity into question; but for the audience, Joan serves as comic relief in an otherwise tense scene. Even with the humor, however, the viewer realizes Joan's difficulty in recognizing God's voice.[32] Joan must learn discernment.

Considering the human form God takes on in every episode, the creators of *Joan of Arcadia* selected an appropriate theme song. "One of Us" (released in 1995) captures the essence of the extra-ordinary depiction of the divine. Performed by Joan Osborne, the song questions God's divine nature by comparing him to a slob and a stranger on the bus. The lyrics convey the notion of encountering the divine in unexpected places and in human form—specifically, as a *stranger*. This metaphor (of the divine as a stranger) may even remind the audience of the scripture: "Do not forget to show hospitality to strangers, for by so doing some people have shown hospitality to angels without knowing it."[33] Other than this allusion to scripture, however, the song suggests a sacrilegious tone, describing God, for example, as a *slob*. While the chorus seems heretical by challenging the divine nature of God, I argue that it is the song's blasphemous claims that make it a fitting choice for a series that pays homage to the historical Jeanne d'Arc. Indeed, not only did God speak to and through the common layperson of Jeanne d'Arc, but historically, this phenomenon became a major concern during her trial. The French maiden was charged and condemned of heresy because she—an ordinary peasant girl—claimed to hear from God. This show insists that God is manifest—acts in and through the world. It is Joan's task to learn to see that and develop her own ability to use both faith and reason to act ethically in the world.

In the end, the overall theology of the series—which explains the manifestations of and responses to God—results from the show's "ten commandments," or Barbara Hall's guidelines for the writers. As previously mentioned, the particular notion that "God talks to everyone all the time in different ways" provides the structural basis for every episode. However, in conjunction with this tenet, Hall offers the principle that "everyone is allowed to say no to God, including Joan."[34] Thus, each character maintains his/her free will when answering the divine call and makes a personal choice as to whether or not to carry out God's mandate. Joan, therefore, uses her own judgment to decide what she will (not) do. While this concept seems quite liberating, Joan and the audience soon realize that her reasoning and judgment cannot grasp the incomprehensible ways of God. Indeed, let us consider Hall's last guideline, which speaks to God's mysterious nature:

> God's purpose for talking to Joan, and to everyone, is to get her (us) to recognize the interconnectedness of all things, i.e. you cannot hurt a person without hurting yourself; all of your actions have consequences; God can be found in the smallest

> actions; God expects us to learn and grow from all our experiences. However, the exact nature of God is a mystery, and the mystery of God can never be solved.[35]

Since God's ways are inscrutable, they must be undertaken with an element of faith and, thus, without full human understanding. The show's theology, then, supports the premise of *éducation* in the medieval sense, as applying to both mind and spirit: Joan responds to each instruction with reasoning (and free will) in addition to faith. Much like the philosophy of Blaise Pascal, introduced earlier in this essay, both concepts—*la raison* and *la foi*—are essential for Joan and must work together.

Conclusion

In the recent trend of "television medievalisms," *Joan of Arcadia* is exceptional as an homage to the Maid that respects the historical Jeanne and proves that the Maid's story is timeless. By revisiting medieval figures and concepts within a modern context, the series brings the past to the present. While other series, like *Reign* (2013) set in France in the mid–1500s, depict fantasies of the past (using some anachronistic storylines and costuming) and give the audience a false sense of "knowing" the Middle Ages, *Joan of Arcadia*, it could be argued, offers a more "authentic" medievalism by promoting Jeanne's notions of faith. The show's theology asserts that it is not unusual to encounter God personally—as Joan does in every episode—and that faith is often a call to action (usually met with opposition).[36]

Finally, if we consider that Howard Bloch and Stephen Nichols describe the field of medieval studies as "a recognition of and insistence upon a certain identity between the medieval period and our own,"[37] it becomes clear that *Joan of Arcadia* epitomizes this aspect of medievalism. By transporting the past to the present, this television series communicates the transmutability of the Middle Ages *par excellence*. In his essay "Modernism and the Politics of Medieval Studies," Nichols addresses the notion of *translatio*: "The Middle Ages had its own concept of transition and transmission which it called *translatio*. But *translatio* was a movement across space and time, a movement of spatial concepts through time which declared themselves in new ways and in new forms (*renovatio*), all the while proclaiming their adherence to the old."[38] *Joan of Arcadia* exemplifies this "movement across space and time" by giving Jeanne d'Arc a new form—*renovatio*—as a 21st-century American teenager who grapples with the need to balance her divine calling with society's expectations or, in other words, faith and reason.

Notes

1. Barbara Hall (the show's creator) lists this idea as one of the writer's guidelines, known as the Ten Commandments of *Joan of Arcadia.*

2. To distinguish between the two heroines, I will henceforth refer to the television character as Joan and the historical figure as Jeanne.

3. See Evelyn Birge Vitz, "Liturgy as Education in the Middle Ages," in *Medieval Education*, ed. R. Begley & J. Koterski (New York: Fordham University Press, 2005), 20 (emphasis added).

4. Vitz, "Liturgy as Education in the Middle Ages," 30.

5. Voltaire also has a lot to say about Jeanne d'Arc specifically. See his controversial mock epic, *La Pucelle d'Orléans* (1762), in which he depicts a tainted Jeanne who loses her virginity.

6. Vitz, "Liturgy as Education in the Middle Ages," 21–3 (emphasis added).

7. Rose Pacatte, "Joan of Arcadia: An Interview with its Catholic Producer," *Saint Anthony Messenger*, March 2005, http://www.americancatholic.org/Messenger/Mar2005/Feature1.asp

8. See Rose Pacatte, "Joan of Arcadia: An Interview with its Catholic Producer" (emphasis added). Also, it is interesting to note that Joan Girardi's younger brother Luke loves science. He is the "science nerd," sometimes overlooked by the family who often focus their attention on Joan or her older brother Kevin.

9. All citations from the show will be referenced in the following manner "1.9" to indicate Season 1, episode 9. Most citations are taken from the online transcript of the episode: http://movie.subtitlr.com/subtitle/show/176116.

10. See Michel Winock, "Joan of Arc," in *Realms of Memory: Rethinking the French Past*, ed. P. Nora, vol. 3 of *Symbols* (New York: Columbia University Press, 1996), 436–437.

11. Even as recently as 2011, a new Johannic film—*Jeanne captive*—debuted in France.

12. See Robin Blaetz, "Joan of Arc and the Cinema," in *Joan of Arc, A Saint for All Reasons: Studies in Myth and Politics*, ed. D. Goy-Blanquet (Burlington, VA: Ashgate, 2003), 164 (emphasis added).

13. See Jeffrey O'Leary, "Joan of Arc: The Maid of Lorraine—Deliverer of Orléans and France," in *The Centurion Principles: Battlefield Lessons for Frontline Leaders* (Nashville: Thomas Nelson, 2004), 73.

14. Aside from the legal sense of the term and its significance in Jeanne's case, even the literal meaning of *habeas corpus* ("have the body") foreshadows her tragic end.

15. The Chambers Brothers originally released this song in 1967, and Joan Jett later covered it in 1990. (Note that the DVD collection released by CBS contains a different song in this scene.)

16. See Karin Beeler, "Teen Visions of God: Postfeminist Heroism and Genre Crossing in *Joan of Arcadia*," in *Seers, Witches, & Psychics on Screen: An Analysis of Women Visionary Characters in Recent Television and Film* (Jefferson, NC: McFarland, 2008), 90 (emphasis added).

17. W.S. Scott, trans., *The Trial of Joan of Arc, Being the verbatim report of the proceedings from the Orleans Manuscript* (International Joan of Arc Society, 1956), http://smu.edu/ijas/1431trial.html/

18. The writers of *Joan of Arcadia* titled another episode "State of Grace" (1.14).

19. Beeler, "Teen Visions of God," 91.

20. Throughout the episode, Joan wears a small silver medallion necklace featuring a profile that resembles an image of Jeanne d'Arc.

21. Beeler, "Teen Visions of God," 90.

22. Even the name *Dreisbach* is quite fitting in a story that focuses on the notion of being bogged down in the mud. One source explains the meaning of the name: "When 'bach' is associated with 'dreis' the resulting meaning is a brook which has its source in a boggy place." "The Dreisbach Name," *The Dreisbach Family Association*, Jan. 24, 2001, http://freepages.genealogy.rootsweb.ancestry.com/~dreisbachfamily/name.htm

23. Ironically, the inscription over the front doors of the high school reads "Enter to learn, go forth for service." This motto is clearly visible in the background at the precise moment when Joan holds up the protest sign and effectively joins the revolution begun by Grace.

24. Vitz, "Liturgy as Education in the Middle Ages," 29.

25. See Angela Zito, "Can Television Mediate Religious Experience? The Theology of *Joan of Arcadia*," in *Religion: Beyond a Concept*, ed. H. de Vries (New York: Fordham University Press, 2008), 729.

26. Zito, "Can Television Mediate Religious Experience?," 737.

27. The visitations from three God-like figures in this episode could be read as another parallel to the historical Jeanne who had three messengers from God: Saints Michael, Catherine, and Margaret.

28. In the script, this character is known as "Domestic Worker God."

29. The mysterious quality of God perhaps alludes to the scripture in Isaiah 55:8–9 ("For my thoughts are not your thoughts, neither are your ways my ways, saith the LORD. For as the heavens are higher than the earth, so are my ways higher than your ways, and my thoughts than your thoughts" [King James Version]).

30. See Rose Pacatte, "Joan of Arcadia: An Interview with its Catholic Producer."

31. Zito, "Can Television Mediate Religious Experience?," 736.

32. Interestingly, the show's first season ends with Joan's hospitalization for Lyme disease and offers a possible explanation for her encounters with God as hallucinations. See episode 23 entitled "Silence."

33. The verse is taken from Hebrews 13:2 (New International Version).

34. In this episode, Joan's own father, for example, tells her that he does not believe in God. When Joan questions him as to what God would have her do, her father counsels her to "figure out how to do the most good" (1.9).

35. Zito, "Can Television Mediate Religious Experience?," 732.

36. On the first point, Evelyn Vitz confirms the belief in God's accessibility: "in medieval, liturgically centered Christianity, the laity experienced, and expected to experience, real closeness to God, the Virgin, and the saints. Such experiences were, to be sure, miraculous; but, as we know, miracles were understood to be anything but uncommon." See "Liturgy as Education in the Middle Ages," 27–8. Secondly, the view of faith leading to an active response was also typical of the Middle Ages. In addition to Jeanne d'Arc, the Crusades provide an earlier example of faith as a call to action/arms in the Middle Ages.

37. See Howard Bloch & Stephen Nichols, "Introduction," in *Medievalism and the Modernist Temper*, ed. H. Bloch & S. Nichols (Baltimore: John Hopkins University Press, 1996), 3.

38. See Stephen Nichols, "Modernism and the Politics of Medieval Studies," in *Medievalism and the Modernist* Temper, ed. H. Bloch & S. Nichols (Baltimore: John Hopkins University Press, 1996), 38–39.

Bibliography

Beeler, K. "Teen Visions of God: Postfeminist Heroism and Genre Crossing in *Joan of Arcadia*." In *Seers, Witches, & Psychics on Screen: An Analysis of Women Visionary Characters in Recent Television and Film*, 83–92. Jefferson, NC: McFarland, 2008.

Blaetz, R. "Joan of Arc and the Cinema." In *Joan of Arc, A Saint for All Reasons: Studies in Myth and Politics*, edited by D. Goy-Blanquet, 143–174. Burlington, VT: Ashgate, 2003.

Bloch, H., & S. Nichols. "Introduction." In *Medievalism and the Modernist Temper*, edited by H. Bloch & S. Nichols, 1–22. Baltimore: John Hopkins University Press, 1996.

Database of Movie Dialogs. "*Joan of Arcadia* (2003): TV series, Season 1, episode 9." Retrieved June 15, 2013. http://movie.subtitlr.com/subtitle/show/176116.

"The Dreisbach Name." *The Dreisbach Family Association.* Jan. 24, 2001.http://freepages.genealogy.rootsweb.ancestry.com/~dreisbachfamily/name.htm.

"Hebrew 13:2." *Bible Gateway.* Retrieved July 31, 2014. https://www.biblegateway.com/passage/?search=hebrews+13%3A2&version=NIV.

Jett, J. *Time Has Come Today*. By The Chambers Brothers. 1968. In *The Hit List* 1990 by Blackheart Records. Compact disc.

Nichols, S. "Modernism and the Politics of Medieval Studies." In *Medievalism and the Modernist Temper*, edited by H. Bloch & S. Nichols, 25–56. Baltimore: John Hopkins University Press, 1996.

O'Leary, J. "Joan of Arc: The Maid of Lorraine—Deliverer of Orléans and France." In *The Centurion Principles: Battlefield Lessons for Frontline Leaders*, 70–98. Nashville: Thomas Nelson, 2004.

Osborne, J. *One of Us*. By E. Bazilian. 1995. In *Relish* by Mercury. Compact disc.
Pacatte, R. "Joan of Arcadia: An Interview with its Catholic Producer." *Saint Anthony Messenger*. Retrieved June 15, 2013 http://www.americancatholic.org/Messenger/Mar2005/Feature1.asp.
Pascal, B. *Pensées. PhiloSophie.* Retrieved February 4, 2014. http://www.ac-grenoble.fr/PhiloSophie/logphil/textes/textesm/pascal1.htm.
Scott, W.S., trans. "The Trial of Joan of Arc, Being the verbatim report of the proceedings from the Orleans Manuscript." *International Joan of Arc Society*. Retrieved June 15, 2013. http://smu.edu/ijas/1431trial.html/.
"St. Joan." Directed by R. Anderson. In *Joan of Arcadia*: The First Season. DVD. Directed by Martha Mitchell. 2003; Sony Pictures & CBS Broadcasting, 2005.
Vitz, E. "Liturgy as Education in the Middle Ages." In *Medieval Education*, edited by R. Begley & J. Koterski, 20–34. New York: Fordham University Press, 2005.
Voltaire. *Dictionnaire Philosophique* "Foi." American Libraries Internet Archive. Retrieved February 4, 2014. https://archive.org/stream/oeuvrescompltes14condgoog#page/n604/mode/2up.
Winock, M. "Joan of Arc." Translated by A. Goldhammer. In *Realms of Memory: Rethinking the French Past*, edited by P. Nora. Vol. 3 of *Symbols*, 433–480. New York: Columbia University Press, 1996.
Zito, A. "Can Television Mediate Religious Experience? The Theology of *Joan of Arcadia*." In *Religion: Beyond a Concept*, edited by H. de Vries, 724–738. New York: Fordham University Press, 2008.

William Webbe's Wench

Henry VIII, History and Popular Culture

Shannon McSheffrey

Henry VIII surely ranks at or near the top of the list of most recognizable kings of the European past. Especially in Holbein's iconic Whitehall mural portrait, Henry has an instant recognition factor that few of his predecessors or successors could claim.[1] We know he had lots of wives and that he had a penchant for cutting off their heads. Thanks to his sturdy stature in Holbein's portrait and the enduringly influential 1933 film version of his life, *The Private Life of Henry VIII* starring Charles Laughton,[2] he is also known as a man who liked his food. From his association with appetites of all kinds have come "Henry VIII Feasts" (where "serving wenches" bring you roast chicken) and a 2007 chocolate bar television commercial featuring diverse people of the past, including the Holbein Henry, feasting on Snickers bars.[3] Until recently, in popular culture Henry was thus almost always a chubby poultry-loving bon-vivant with a bad track record with wives. He was of course a natural, then, for a 2004 episode of *The Simpsons*, where Homer fulfills all the clichés, eating compulsively and beheading all his queens, except for his first, blue-haired wife, Margarine of Aragon.[4] The creators of the recent television series *The Tudors* thus cast deliberately against type when they chose a conspicuously fit and slim Jonathan Rhys Meyers to play Henry VIII.[5]

The Tudors, a Canada-Ireland co-production made primarily for the Showtime cable network in the U.S. and broadcast by the CBC in Canada, BBC2 in Britain, and TV3 in Ireland, was created, written, and executive-produced by the British television and film writer Michael Hirst. Hirst is known for his historical dramas; he wrote the two *Elizabeth* films starring Cate Blanchett and a number of other historical films and television series, including most recently *The Borgias* and *Vikings*.[6] His version of Henry VIII's court focuses much more on the king's reputation for sexual voraciousness than on his love of chicken. Unlike the Charles Laughton or Homer Simpson versions, this Henry is not at all comic, or at least never intentionally so; the series focuses on a heady mixture of sex, religion, and politics. Henry and many of his closest companions are played by actors far younger than the actual historical figures, and (unsurprisingly) much better looking. Hirst evidently was interested in using the historical setting and relatively well-known dramatic events of Henry VIII's court and life—his many marriages, the English church's split from Rome—to dramatize the conflicts and moral dilemmas faced by those born with beauty, power, and wealth. Few would argue that the series advances our historical understanding of the period, and of course it was not intended as a piece of scholarship. Whether the series succeeds as television is a matter of taste; it was mostly panned by television critics but has a large and loyal fan base.[7] *The Tudors*' Facebook page surpassed the 1 million "likes" mark in May 2013 and continues to be very active although the last episode aired in 2010.[8]

The idea of Henry's court as a playground for the young, well-born, and beautiful—and the idea of Henry himself as strapping, handsome, and athletic—is not entirely fanciful. Henry acceded to the throne when he was eighteen. In his youth Henry was tall, slim and well-built, and at his court, surrounded by companions of his own age and a beautiful queen in her mid-twenties, he delighted in sponsoring masques, revels, feasts, and jousting tournaments.[9] Hirst's approach is to telescope Henry's reign, to imagine the young Henry living the *Entourage* lifestyle at the same time as he implemented major changes to state and church. As Hirst has said, the conceit for the series is that "the courts of Europe were run by people in their teens and twenties.... That's why they were so crazy. We have this image now that the court is always middle-aged, but it wasn't true."[10] Hirst is, of course, right; kings were sometimes young and handsome, and the series successfully captures a moment around 1520 when the kings of Spain, France, and England were all young, athletic, good-looking, intellectual, and ambitious men. But just as Henry's court was not always middle-aged, it was also not forever young—and the Master Narrative moments of Henry's reign (divorce, Reformation, and five of Henry's six marriages) occurred when Henry was over forty, and no longer slim, athletic, and good-looking. There would be no point, however, in simply focusing on the early part of the reign when the *Entourage* comparison would be more appropriate, because those years have none of the well-known events that viewers associate with Henry.

The Tudors is among a recent wave of intended-for-American-cable series that might best be called, to use critic Mike Hale's apt term, "frequently-removed-costume dramas"[11]: they feature graphic violence and as much nudity and sex as television standards will allow. *The Tudors*, inspired by historical dramas on HBO such as *Deadwood* and *Rome*, was a commercial success, and so in turn it begot other historical television series on American cable networks: *The Pillars of the Earth*, *Camelot*, *World Without End*, *The White Queen*, *The Borgias*, and *Vikings* (the latter two involving the same production team as *The Tudors*). The other series, save for the Renaissance-themed *Borgias*, all have medieval settings, and in many ways *The Tudors* fits happily within the phenomenon of television medievalism. Henry VIII's reign—more so than that of his daughter Elizabeth, also of course a frequent subject for popular media forms—carries many of the same signifiers as the Middle Ages in modern Western culture. Like the medieval period, the Henrician age is pre-industrial, bucolic, and "traditional." Violence is frequent, bloody, and too often unjust; socio-political hierarchies are rigid, kings often despotic; people live in castles and manor houses; costumes are sumptuous; ladies are beautiful and aristocratic men look chivalric and virile whether in their armor or in their hose and high leather boots.

Yet *The Tudors* does not quite fit the medievalism paradigm, and indeed is the best-known recent example of the cultural phenomenon scholars have labeled "Tudorism," a cousin of the much-better-studied medievalism.[12] In particular, while the Middle Ages stand in popular culture as the obverse face of modernity, both in its embrace of "traditional" values of honor, bravery, and chivalry and in its rejection of technology, progress, and rationality, the Tudor age by contrast is characterized as standing on the threshold of the modern age. Like other Tudorist products, *The Tudors* presents to its viewers the attractions both of castles, armor, and velvet gowns, and the sense that its characters—who see the Reformation, New World explorations, and technological advance—live at the dawn of "us."

The particular popular interest in Henry VIII also means that *The Tudors* centers on a figure who is much better known to audiences than perhaps any historical medieval figure. Probably for as simple a reason as his colorful marital history, Henry is memorable. The creators of *The Tudors* thus had both challenges and opportunities posed by audience expectations for Henry VIII that historical characters in the recent medieval-set television dramas, such as Queen Maud, Edward III, Edward IV, or Rodrigo Borgia, did not present. The challenges and opportunities inherent in depicting historical periods that readers and listeners already think they know are very familiar to academic historians, who face these issues routinely in the classroom and in any writing they do for students or general readers. This essay will both consider how the creators of

The Tudors navigated what viewers knew about and expected to see regarding Henry, and reflect on the differences between what scholars and scriptwriters make of the remnants of the past with which both work.

"Accuracy" and Screen History

Does it matter that Jonathan Rhys Meyers does not look like Henry VIII? Conventionally, historians have decried the inaccuracies in historical films and television programs, acting as "historian-cops" in pointing out the errors, big and small, in screen treatments of the past. Scholars who study the presentation of the past on screen have urged us to move beyond such "fidelity criticism," arguing that veracity is beside the point in historical films and television—that historians who hunt out "errors" have misunderstood the ways the past is used in popular culture and have emphasized more than do most working academic historians a positivist approach to the knowability of the past.[13] Jerome de Groot has argued in particular regarding *The Tudors* that the series deliberately disavows any attempt to know the actuality of history. Hirst, he contends, uses the series to deride academic history as "creaky, unsexy, and ineffectual," consciously playing with historical tropes about Henry both in order to challenge popular cliché and to argue that "all historical representation is merely a recapitulation of something unknowable," challenging "the reality of the official discourse of the past."[14] *The Tudors*, he contends, makes no attempt to be true to the historical record and to assess it on those grounds is to miss its point.

Some of the "inaccuracies" in *The Tudors* do indeed knowingly play with audience expectations, especially the most obvious disjuncture between "fact" and *The Tudors*—everyone knows what Henry VIII looked like, and that he looked nothing like Jonathan Rhys Meyers.[15] Rhys Meyers's anti–Henry was part of the show's schtick: a good deal of the publicity surrounded Rhys Meyers's not being the corpulent ginger-haired man you expect. But if *The Tudors* is in some ways iconoclastic, "a mewling, brattish, present media form" that emphasizes "flash, superficial beauty and surface," the series does not embrace a post-modern rejection of "historical accuracy" to nearly the extent that de Groot claims[16]—in fact it uses claims to the "historical truth" of what appears on screen as an important element of its appeal to viewers.

As Richard Burt has argued, turning away from "fidelity" has allowed us to understand more fully the uses of the past in modern media, but it has prevented serious discussion of an important aspect of those cultural products: their invocation of a real, authentic past.[17] No doubt to some viewers the question of "accuracy" is irrelevant—both for those who like the show and for those who don't.[18] For many others, however, historical television and film dramas attract precisely because they are about a "real" history, and the creators of historical dramas are very attentive to those desires in their viewers. Although the purpose of a show like *The Tudors* is not to present scholarship but to entertain, to give the viewer pleasure, the entertainment value of what happens on the screen cannot be entirely divorced from its historicity. Historical dramas draw on the attractions and the pleasure of narratives "based on a true story"; whether or not the story told actually bears a close relationship to "what really happened," clearly the claim of authenticity is thought to resonate with audiences.[19]

In the publicity that accompanies historical television series and films, directors, producers, writers, and actors repeatedly invoke the copious amounts of "historical research" that underpins what appears on the screen.[20] As Burt has commented, new media forms have only increased the resources film and television producers provide viewers. Paratexts that accompany the depiction on screen—DVD features, commentaries, onscreen pop-up bubbles, apps, Facebook groups, Twitter feeds, and so on—often focus on the "real history" that lies behind the scenes. *The Tudors* are no different; in a DVD feature for Season 2, for instance, entitled "The Tower of London," Tom Stammers, identified as "*Tudors* historian"[21] (not Tudor historian, but *Tudors* historian), leads Natalie Dormer, the actress who plays Anne Boleyn, around the Tower to visit the site of Anne's final days. Repeatedly the dialogue emphasizes the historicity of the scenes in the program: "How historically accurate is that?" Dormer asks about a particular plot point; "Undoubtedly accurate," Stammers replies. Presumably this rhetoric of authenticity and truth is so frequently used because it works as a marketing tool.

The Tudors writer, Michael Hirst, has emphasized the substantial research he conducted for the series's scripts. Although he concedes that he played around with chronology, Hirst claims that the show is "85% historically accurate"; Hirst in fact says that he created *The Tudors* to "correct" the public's perception of Henry as "he appears in the famous Holbein portrait, all ruffs around his neck and bulging belly."[22] He also emphasizes his avid reading of works of history and his fascination with footnotes ("I get a lot of juice out of the footnotes in history books").[23] A fan page for *The Tudors* has established a wiki that links the scenes in the show to the historical works that verify their "truth," complete with a quotation from Michael Hirst at the top which again emphasizes his reliance especially on footnotes "in very dry and learned texts."[24] Other historical television series and films have similarly emphasized deep research in dusty tomes or even original archival documents: a "making of" feature on the DVD of Luc Besson's Joan of Arc film *The Messenger* claims, for instance, that the film "was based on six months of extensive research in the archives."[25]

Many of these projects engage historical advisors, which often represents a rhetorical claim to scholarship and accuracy more than it indicates any significant academic influence on the products. Some of the historical consultants are professional "providers of historical content," as is Justin Pollard, named as historical consultant on *The Tudors* (his many and diverse credits include the films *Pirates of the Caribbean 4* and *Les Misérables*, and the television series *Geldof in Africa* and *Egypt's Golden Empire*).[26] Others are academics; a University of Warwick art historian, Jenny Alexander, appears for instance as historical consultant for the loosely historical *Pillars of the Earth*.[27] As one documentary-maker has put it, these academic advisors are employed primarily as "a public relations gambit" and rarely have any "meaningful input" into the projects, even in documentaries (presumably even less so for historical dramas).[28]

The rhetoric of copious research and accuracy often places those doing publicity for the films or programs in a double bind—because (as those historian-cops have frequently complained) historical dramas on film and television almost always subordinate "accuracy" to the narrative demands of the story on screen. This forces those writers and directors into making internally contradictory statements: the film or program is "all true," but it's a fiction so we've taken liberties; it's accurate, but historians cannot agree anyway and so our interpretation is as good as theirs; it's based on rigorous research, but it's only a movie.[29] As Hirst puts it:

> As a whole, *The Tudors* is remarkably true, but it's drama, not history. You have to condense things and highlight things, but my only resources are books by historians, which I read avidly. All historians disagree with one another anyway, so the idea that there is one forensic truth is itself a lie.[30]

Despite reading historians' works (and especially their footnotes) "avidly," Hirst questions whether they in fact know anything at all:

> It's not like any of the historians were actually there. So what you read in history books, is that historically accurate? Not necessarily. And in any case I'm not writing a documentary.[31]

Although Hirst and others involved in these productions are happy to allude to postmodern challenges to "truth" in history when it suits their purposes, fundamentally their approach to "historical accuracy" is firmly positivist.[32] Although Hirst on the one hand invokes a right to speculate on the points where "facts" are not clear, he maintains that what he has done is as true to the historical record as it can be—until, that is, someone points out that Henry VIII had two sisters rather than the one composite character who appears in *The Tudors*, at which point "creative license" is invoked.[33]

This is not to say that claims of research are entirely spurious—and this is what makes *The Tudors* a particularly curious historical drama. Unlike most other recent historical costume dramas, which have been based on popular fiction (such as *The Pillars of the Earth* and *World Without End*, both based on

Ken Follett novels; *The White Queen*, based on several Philippa Gregory novels; *Game of Thrones*, based on the G. R. R. Martin series), Hirst clearly does use "history books" in writing his scripts. He has not simply designed the dramatic arc from a schematic outline of Henry's reign, as de Groot suggests he does. Almost all the characters, major and minor, are grounded on historical persons, or at least bear their names, and scenes are often inspired by, if not precisely based on, the historical record.[34] Watching the series while working on a project involving reading archival manuscript sources from Henry's reign, on several occasions I had the strange sensation of seeing acted out, in a scene from *The Tudors*, an obscure scenario that I had just read about in an archival document—often wrenched out of its context, but nonetheless recognizably linked to the original evidence. In episode three of the second season, for instance, a sword fight between two courtiers that ended in the death of one of Charles Brandon's retinue was the subject of a brief scene, used by Hirst to show the spiraling quarrel at court between Charles Brandon, the duke of Suffolk, and the Boleyn family. The original records of the subsequent homicide case at the court of King's Bench show a much more complicated situation, which (understandably enough) the three-minute scene does not explore.[35] Hirst employs these kinds of obscure vignettes not because the viewer would expect them—indeed few viewers would be able to distinguish the "strange but true" stories (as *The Tudors Wiki* puts it) from those that Hirst invented—but primarily because, as Hirst himself says, sometimes the stories are so good that he couldn't *not* use them.

Where, precisely, did Hirst find these stories? How did Hirst employ historical scholarship for his conceptualization of the series and for the details of his scripts? What is the relationship between the flourishing industry of history publications on the reign of Henry VIII—from *über*-scholarly journal articles to accessible academic books to populist histories—and his portrayal in 21st-century film and television? To explore these issues, I will look more closely at a short scene from the second season of *The Tudors* that corresponds to a little-studied set of records from the 1530s, dealing with a man named William Webbe and his wench.

William Webbe and His Wench in the Historical Record

First, the 1530s version. I found the documents, a set of files relating to an informal enquiry undertaken at chief minister Thomas Cromwell's behest, in the voluminous set of records in the National Archives known as the State Papers. I came across them while researching the subject of sanctuary in the late medieval and Tudor period.[36] Sanctuaries were small territories attached to religious houses where accused criminals, debtors, and illegal foreign workers could seek asylum. Neither royal nor local civic officials could arrest those who had been given the privilege of sanctuary. Several of these sanctuary precincts, in which rental housing was built for those who wanted or needed to live within the bounds, became communities in themselves, with some people living in them for years and even decades. Perhaps the most important sanctuary in the realm was at Westminster Abbey, which was located right next to the royal palace at Westminster.[37]

On 9 September 1537, Thomas Cromwell, then titled "Lord Privy Seal" and effectively second-in-command to the king, received a letter from a man named Harry Atkinson, who was imprisoned in the "convict house," or prison, inside the Westminster sanctuary. Atkinson, a friend of one of Cromwell's underlings, wrote to the Lord Privy Seal to explain that William Webbe, the keeper of Westminster sanctuary, had unjustly imprisoned him; he hoped that Cromwell would order him to be released.[38] Atkinson explained in his letter that he had expressed concern about a scurrilous story regarding the king that was spreading through the sanctuary precinct, for he knew that this story constituted treason. The story also involved Webbe, and Webbe had, in a rage, thrown Atkinson in prison. As a result of Atkinson's letter, Cromwell mandated a semi-official enquiry to report to him. Witnesses appearing before the enquiry each testified that he had heard, at third- or fourth-hand (never from Webbe himself), that William Webbe had been going around the precinct telling the following story: Webbe had been riding a fair gelding, near one of the royal palaces southeast of London, with "a pretty wench" (never named) behind him. The king met up with them on the road and he said to Webbe, "Webbe, thou art never without such pretty carriage behind thee." Webbe answered that she was a pretty piece for a poor man to pass the time with, and the king "plucked down her muffler," kissed her, and commanded her to alight from the horse. He then took her off and "had his pleasure of her" (the wench's response—whether this was at her "pleasure"—remaining unstated, irrelevant to the men who told and heard this story). The king thus took the woman away from Webbe, who had "kept" her for two years before this. In some versions witnesses told, Webbe swore "a vengeance on him for taking away of my wench." Most of those testifying to the enquiry ended the story with the conclusion, "and thus the king lived in adultery."[39]

This is a colorful and interesting story, but Cromwell and his enquirers apparently decided not only that the encounter it describes between Webbe, the wench, and the king had not actually happened, but also that William Webbe had never told the story in the first place. Although no judgment is recorded—it was an unofficial commission of enquiry and any decision was likely taken informally by Cromwell, perhaps with the king's input—we can infer from a letter to Cromwell from one of the commissioners that the story's genesis was attributed to a certain Robert Sharpe, who had devised the tale for his own malicious purposes.[40] Sharpe coveted William Webbe's job. He hoped that by telling everyone that Webbe was accusing the king of adultery that Webbe would come under suspicion of treason. This would neatly, Sharpe hoped, remove Webbe from his position in the sanctuary. Sharpe's tactic, however, did not succeed; Webbe emerged from this episode unscathed, continuing to govern the Westminster sanctuary for years thereafter (Sharpe's and Atkinson's subsequent careers are unknown).[41] Although we cannot simply accept the judgment of these kinds of ad hoc political enquiries as established fact, in this case the commissioners' judgment seems the most likely scenario. The evidence thus points to the tale of William Webbe's wench as a fiction rather than as a real episode in Henry VIII's life; even the association of William Webbe himself with the story appears to have been invented. It was an interesting tale, though, and it is not too surprising that the enquiry's evidence indicates that it flew easily, through supper-table talk and social chit-chat, among sanctuary men, sanctuary functionaries, visitors to the precinct, servants to the king, and beyond the sanctuary among the inhabitants of the town of Westminster.

If the story's spread does not surprise us, some other aspects of it might, given what we think we know about Henry VIII. The fundamental point of the story, in all the versions the witnesses told, was that the king "lived in adultery." This suggests that Henry was not, as we might have assumed, well-known as a philanderer, but that even among criminals living in sanctuary such a story had the power to shock. The tale could be used to smear William Webbe because to accuse the king of adultery was tantamount to treason. We must recall: this was a delicate time. The dinner party at which Atkinson said he first heard the gossip was in early September 1537 and the inquiry held in the later part of that same month; according to other evidence, the story may have originated in the fall of 1536. The story and its lesson, that the king was an adulterer, was presumably meant, and taken, as an implicit criticism of the grounds for the execution of Anne Boleyn, the king's second wife. Anne was executed for adultery and incest in May 1536 and the king married his third wife, Jane Seymour, within days. At the time of this inquiry in September 1537, Queen Jane was heavily pregnant with the future Edward VI, who was born about two weeks later. Probably also relevant is the episode's timing in the midst of a period of profound religious change; Henry's role as supreme head of the newly formed Church of England added another layer to the accusation of sin. All the witnesses were keen to dissociate themselves from the treasonous statements, each emphasizing that he would never have told the story, that he had only heard it. A palpable anxiety in the documents suggests fear that even hearing the tale might lead to the noose or the even more terrifying execution of drawing and quartering meted out to traitors.

As historical evidence, I would suggest that these records tell us about the role of rumor in the 1530s and subterranean antagonism to Henry's policies; they tell us about the informal and ad hoc processes through which many legal or quasi-legal issues were handled in the 1530s, skirting formal and public legal prosecution; they tell us about everyday life and local politics in the Westminster sanctuary; they tell us about the purely instrumental role women labeled "wenches" were seen to have in sexual transactions by men in and around the Westminster court; they tell us about Henry's sensitivities to talk about his sex life and the complicated relationship of the king's body to the body politic (about which more below). I do not think they tell us about a real episode where William Webbe and his never-named wench met up with Henry VIII on a road near one of his royal palaces, although of course one never knows.

William Webbe and His Wench in The Tudors

Now for the version of the story of William Webbe's wench in *The Tudors*. The scene occurs in episode five of the second season. Rather than occurring in 1537, it is placed in about 1535, at a point where Henry's marriage to Anne Boleyn has entered a rocky stage. After having given birth in 1533 to a girl, the future queen Elizabeth, Anne has just miscarried a second pregnancy, a boy. Sir Thomas More and Bishop John Fisher languish in the Tower, awaiting execution for refusing to recognize the king's supreme headship of the new Church of England. The king's former brother-in-law and closest friend Charles Brandon, duke of Suffolk, who opposed the Boleyn marriage and was exiled from court for it, has recently been re-admitted to the king's circle.

In this scene, Henry and Brandon are riding through a forest and Henry ruminates on an issue that previous scenes make clear he associates with Anne's recent miscarriage, which he fears was caused by her rumored promiscuity before their marriage. Henry asks Brandon: "Have any of the women you've ever bedded lied about their virginity?" Brandon laughs and answers, "I'd say it's the other way 'round: did any of them *not* lie about it." Looking at Henry's face, he realizes that he has said the wrong thing, and begs Henry's forgiveness; Henry says, "It doesn't matter. I asked for the truth and you told me."

At this moment they encounter a man and a woman riding on the forest road toward them. The guards bark at them to dismount, and Henry, also dismounting, approaches them as they bow and curtsey deeply before him. The man is dressed in non-descript commoner clothing, and the woman wears a long riding cloak over a simple gown, her head wrapped in a shawl. We cannot see her face, as her eyes are demurely cast down.

"Good morrow, Your Majesty," the man says.

"Good morrow," Henry replies. "What's your name?"

"William W-Webbe, Your Majesty," the man stutters.

"No. Your sweetheart's name," Henry says.

"Bess, Your Majesty," Webbe answers, looking mystified.

Henry approaches the woman, saying, "Come here, Bess," as he bids her rise from her deep curtsey.

As Henry unwraps the shawl from Bess's head, Webbe babbles, "I assure your majesty I have a permit and permission to ride through your forest; I swear it and can easily prove it."

Henry, however, pays no attention as he examines Bess's face (she's remarkably beautiful) and then kisses her. He says, "Come with me," and leads Bess, who is now smiling, to his horse. Webbe is left on the road, trying not to look affronted, while Brandon smirks.

Cut to a big four-poster bed in a royal bedchamber, where Henry and Bess are having passionate sex. "Are you really the King of England?" Bess asks.

"No," replies Henry, "I was only pulling your leg." They climax noisily.[42] Neither Bess nor William Webbe appear again in the series.

Both the *Tudors* version and the Westminster sanctuary gossip version are stories. What is interesting here is looking at the differences in how they are told, and the points they are meant to impart. In the modern version, the story is used to illustrate Henry's freewheeling sexual appetites; not surprisingly given the narrative logic of the series, it is read straight ahead, jumping straight to the purported scene rather than considering it as an unsubstantiated rumor. Unlike the 1537 version, the wench has a name—Bess—and she clearly is pleased by the encounter. The scene fits into the main narrative lines of the episode: it illustrates Henry's doubts about Anne's chastity before they married and their growing estrangement. It also suggests Brandon's satisfaction that Henry has begun to turn away from Anne, which he hopes will result in a shift in the balance of power on the king's council away from the Boleyns and toward him. It also offers an opportunity to provide a sex scene, without which no episode of *The Tudors* would be complete.

Television History and Source Material

Thus, not too shockingly, the version of a historical moment told in *The Tudors* strips off the complications, places it in a different time, and focuses on the sex. My surprise in seeing this scene play out on the television screen was not in the way it was used in the series, but that it was there at all. So where did Hirst find the William Webbe episode? It turns up, as far as I am able to determine, in only one scholarly book on Henry VIII's reign, G. R. Elton's 1972 *Policy and Police*, where Elton (the most prominent Tudor historian of his generation) interprets it in much the same way I have, treating it as an example of the kinds of rumors that were current in the period after Anne Boleyn's execution, and as evidence for Henry's sensitivity particularly about his and his queens' sexual reputations. He does not treat it as an actual event.[43] If this were where Hirst found this episode, he was thus choosing to interpret it rather loosely. I am fairly sure, however, that Hirst did not use Elton's book, and that his much-vaunted historical research emphasizes popular histories rather than the "very dry and learned texts" he says he consulted.

The *Tudors* wiki on Hirst's historical sources includes an entry on the William Webbe episode, and identifies, correctly I believe, the source: Alison Weir's *Henry VIII: The King and His Court*, published in 2001.[44] Alison Weir, a prolific writer of popular history, has written a very readable biography of the king, based largely on published primary sources. Her version of the episode, as the wiki indicates, is not complicated:

> In the late 1530s, a man called William Webbe complained that, whilst he was riding in broad daylight with his mistress near Eltham Palace, they encountered the King, who took an immediate fancy to the "pretty wench," pulled her up onto his horse, and rode with her to the palace where he ravished her and kept her for some time.[45]

Note here that William Webbe is just a man; he is not keeper of the Westminster sanctuary personally familiar to the king, which was a crucial part of the situation as outlined in the actual documents, and the sanctuary itself is wiped out of the telling. Weir here uses the word "ravish," which connoted abduction and rape in the 16th century, although possibly she means it in a looser sense simply to indicate sexual intercourse; the original sources do not discuss the wench's agency at all in the episode, but nor do they suggest that the king took her by force. The mediated nature of the tale—the fact that the evidence is *not* a complaint from William Webbe directly, but rather a third-hand rumor that he had complained—is entirely lost, not to mention the examiners' evident conclusion that the story had been invented in the first place. Weir not only treats it as a fact that Webbe complained, but also treats as factual the subject of the complaint, Henry's taking of the wench. Weir relates the incident as one of a number of pieces of evidence that show (she argues) that Henry still had a "wandering eye" even in his later years and was known to have had many casual sexual encounters. Weir herself did not read the original documents in the case, but used the published summaries of the archival material in the multi-volume *Letters and Papers of Henry VIII*. Some of Weir's misconstruing of the case is due to the incomplete and (uncharacteristically) misleading summary of the original documents in the *Letters and Papers*.[46] She nonetheless has taken even that material out of context, focusing on the interesting tale without accounting for why such a story was recorded in the first place.

Apart from the dating, Weir's text is clearly the germ of the scene that appears in *The Tudors*; the wresting of the narrative from its context was Weir's work, not Hirst's, and in this case he is actually very faithful to his source, with one obvious difference—the meaning of the sexual encounter, as titillating or distasteful. That meaning is determined largely by the divergent characters of the king in the two narratives—the story of the king taking William Webbe's wench has an entirely different valence when the hunter is young, sexy, and fit, than when he is overweight and old before his time. In the TV series, Henry is played by a slim, good-looking actor, who was about thirty when the second season was shot. The real Henry about whom the sanctuary men told the stories in 1537, and about whom Weir wrote in her telescoped version of the story, was a very large and unfit 46-year-old.

The famous Holbein portrait dates from that same year, but this is likely an idealized version, using perspective to make the king look sturdy rather than seriously overweight. Following a serious accident at a tournament in 1536, Henry's health seems to have taken a turn for the worse and he became much less active than he had been, with resulting weight gain. The king's armor in the late 1530s indicates he would have taken an XXXL in today's sizes[47]; a tall man of between 6'2" and 6'4", at that time he probably weighed around three hundred pounds. Weir uses the story to support the general point she makes in that section of her book, that despite his physical deterioration, Henry remained sexually rapacious. By contrast, clearly Hirst had no interest in portraying Henry as a fat and lecherous middle-aged man; indeed he and others explicitly linked the importance of having an attractive actor playing Henry to the abundant sex on screen. As Rhys Meyers himself put it in an interview with the *Daily Mail* in 2009, "The reality is that viewers don't want to see an obese, red-haired guy on a TV series. I mean, I wouldn't like to see somebody who looked like Henry when he was older having sex."[48]

If Hirst relied closely on Weir's biography of Henry as he wrote his scripts, her book is hardly the kind of "very dry and learned text" Hirst describes himself eagerly reading, and by no means could he have spent much time looking at her footnotes, which are highly uninformative.[49] This is not to say that Hirst *should* have used scholarly history with lots of boring footnotes: filming my version of the story of William Webbe's wench, with its required convoluted explanations about sanctuary, its insistence on the third-handedness of the tale, and its conclusion that the scenario itself probably did not happen, would be long, complicated, and perhaps not all that entertaining. It might make an interesting postmodern film that subordinates a straightforward storytelling to multiple viewpoints and an unstable narrative, but it would not suit the kind of television costume drama Hirst devised in *The Tudors*. However, Weir's brief vignette translates well into a quick punchy scene that is perfect for *The Tudors*, moving the plot along nicely.[50]

Sex, Henry VIII and The Tudors

Weir's version of William Webbe's wench also works better than the scholarly version because—despite the series' stated determination to challenge the stereotypes about Henry VIII—it comfortably confirms what we already "know" about the king. Henry had a lot of wives, and thus must have been sex-mad, and in any case there were lots of bosoms and wenches back in those days. But do we really "know" this about Henry VIII?

Although in *The Tudors*, the rampant sexuality of the king and his courtiers is open and unconstrained, in real life Henry's court was not a sexual playground. Nor, even in private, was his sex life anywhere near as wanton as the modern TV version, despite the multiplicity of wives. Precisely how sexually active Henry was remains a matter of scholarly debate.[51] Historians agree that he was not altogether faithful to Katherine of Aragon; he had one acknowledged illegitimate child, Henry FitzRoy, the earl of Richmond, born in 1519, and two known mistresses, Elizabeth Blount, Henry FitzRoy's mother, and (somewhat ironically) Mary Boleyn, Anne's sister, with whom he had a relationship in the early 1520s. As he entered the more eventful parts of his reign—the mid–1520s, 1530s, and 1540s, the time period the TV series covers—the evidence that Henry had partners outside of marriage becomes ambiguous. The rumors and stories that circulated in the 1530s, of which the tale of William Webbe's wench is but one example, have led some scholars to contend that Henry was promiscuous. Those rumors, however, are far from smoking guns, and most historians argue that Henry's sexual conduct after the mid–1520s was unadventurous and even perhaps entirely confined to the marriage bed.[52] Indeed, his biographer in the *Oxford Dictionary of National Biography*, Eric Ives, states that by the 1530s Henry was "certainly having psychosexual problems," and may even have been intermittently impotent.[53] If Henry was seeking sex outside of his marriage, he did not do so openly nor (as we have seen) did he tolerate speculation about it.

This is not to say that Henry's sex life, and his fertility, were issues of only private concern. The king's body in both figurative and literal senses was the body politic, and his generative sex acts were deeply important to the future of his kingdom—perhaps particularly as he lacked (until September 1537) a male heir. If Henry's advisors sought to suppress any suggestion of the king's adultery, at the same time they promoted through official portraiture Henry's fecundity and, necessarily in tandem, his sexuality. Both his fertility and his potency were implicitly in question by 1536 if not before, perhaps in his own mind, as Ives suggests, as much as in whispers among his subjects. His queens repeatedly miscarried and produced only girls, and doubts mounted as his health took a turn for the worse after the 1536 accident. The portraits of the mid- and late-1530s, Kevin Sharpe argues, thus constituted a public relations offensive, designed to counter any doubts about the future of the regime. As Sharpe comments, the famous Holbein Privy Chamber portrait, dating from 1537 before his son's Edward's birth, is "priapic," leaving no doubt that the lack of a male heir thus far had by no means been due to any deficits on his part.[54]

If sometimes we feel that *The Tudors* emphasizes sexuality too much, it should be noted that not even that television series emphasizes the male genitals as much as the prominent codpieces in Henry's actual portraits, and his armor, did.[55] One distinct difference between Henry's actual dress—at least in his portraits—and Jonathan Rhys Meyers's costume in the television series is the latter's *de*-emphasis of genitals. Indeed comparing Holbein's portrait of Henry with a standard publicity still for the series where Rhys Meyers poses in a similar stance, Rhys Meyers seems almost neutered.[56] The prominence of Henry's codpiece serves not only to highlight the ambiguities of Henry's public image in the 1530s, but also shows up a contemporary tendency both to avoid explicit references to penises in many cultural representations of masculine potencies, and to de-link priapism from the issue of fertility.

The portrayal of sex in *The Tudors* is one of the most obvious distinctions between this "Tudorist" television series and other recent television historical dramas set in the medieval period. To be sure, as in most medieval series, the plot-lines in *The Tudors* assume that high-status men could take women (or, occasionally, men) at will, the plucking of sexual fruit assumed as an element of aristocratic or royal privilege.[57] The use of sex by the men of the royal court in *The Tudors*, however, is somewhat different from the popular view of aristocratic sexual rapacity in the Middle Ages, where kings and nobles seize women, especially those of lower rank, without regard for their consent. The 1995 film *Braveheart* revived in the Anglophone world a centuries-old myth about medieval feudalism, that there existed a "right of the first night" [*jus primae noctis*], by which medieval lords customarily had the entitlement, as part of a feudal lord's rule over his dependents, to deflower the brides of their dependents on the night of the wedding.[58] In *Braveheart*, an English lord's determination to take his "*prima nocta*" with a local Scottish bride ignites a Scottish rebellion of independence, the lord's taking of the bride's body clearly serving as a metaphor for England's rape of Scotland.[59] In the recent spate of medieval television productions, the plot-lines strongly associate aristocratic male sexuality (and in some cases, any kind of medieval male sexuality) with coercion. In *The White Queen*, which chronicles the Wars of the Roses in 15th-century England, most of the sexual relationships in the dynastic marriages are brutish couplings in which husbands force themselves on their frightened wives; even the central and erotically charged relationship between King Edward and his future wife Elizabeth Woodville begins with his nearly raping her.[60] One of the central turns in the plot of *The Pillars of the Earth* occurs when the earl of Shiring's beautiful daughter, Lady Aliena, is raped by the dastardly William Hamleigh, whose father has usurped the earldom.[61] In *The Vikings*, scenes with female characters, especially in early episodes, seem more often than not to involve repelling rapists, a gesture to contemporary girl-power framed in a medieval world of sex as rape.[62] *World Without End* takes the sexual violence furthest, with perhaps more scenes of rape than consensual sex.[63]

If most of the recent medieval television series employ a brutalist vision of sex before the civilizing force of modernity, *The Tudors* presents a model of erotic relationships that emphasizes sex as strategy and commodity—a paradigm perfectly in line with modern tales of sex and the single girl from Helen Gurley Brown through *Sex and the City* to *The Bachelorette*. Coercive sex is mostly absent from *The Tudors*; although during the last season and a half the sex turns darker and less fun (including two rape scenes),[64] in the first two seasons, sexual connections on *The Tudors* are free from any hint of force. In contrast to the medieval series—including *The Vikings*, another Hirst vehicle—the women in *The Tudors* do not have to suffer seemingly constant brutal and unwanted sexual advances; instead they are willing partners of men who seduce them, as is William Webbe's Bess, who does not seem at all displeased to be taken by Henry. The tone is set in the first episode, when Henry, bedding his second lady-in-waiting of the hour, prefaces their lovemaking by asking, "Do you consent?" Already naked, she unironically signals both her desire and her recognition of his sovereign status by breathily answering, "Yes, Your Majesty." Yet even if his royal status commands, it is not simply Henry's majesty that attracts women, but his beauty: as Ramona Wray has remarked, "Henry's desirability cuts across any question of consent." What woman would refuse him as bedpartner, king or not?[65]

Not all women, nor all men, are part of this sexual economy in *The Tudors*; there is no suggestion that Sir Thomas More's eye wandered, for instance, nor that his wife or daughter would either excite or accept sexual overtures outside marriage. This is not problematic in the series: those characters signal by the sobriety of their costume and facial expressions that such things are unthinkable, and thus they simply do not arise in the plot. Women whom powerful men might want sexually are marked by their demeanor, physical beauty, and cleavage, and they are willing when asked.[66] The female sexual partners gain, too; the king's attention is in itself a reward for their beauty, and implicitly and sometimes explicitly they are also rewarded materially. If the series assumes that powerful men, especially the king, could have whomever they pleased, they (mostly) want only those who are happy to acquiesce. Anne Boleyn's famous story—as the woman who won the crown by being the only one who held out—drives much of the plot of the first two seasons, but it too ultimately confirms the same assumptions about women and the strategic uses of sex in the series.

Conclusion

Michael Hirst sought material and inspiration for *The Tudors* from "what really happened"—both the obvious (the divorce; Anne Boleyn's beheading; the split from Rome) and the relatively obscure (the dispute that led to the slaying of one of Charles Brandon's men; William Webbe's wench). Hirst shaped both the major familiar events and the little-known vignettes (which almost no one in the audience would be able to distinguish from the scenes he had invented from whole cloth) to suit his narrative requirements. Although in publicity for *The Tudors* Hirst has emphasized his deep research in obscure works of historical scholarship—dusty tomes with many footnotes—his sources for those vignettes were instead popular biographies of Henry VIII. As in the case of William Webbe's wench, the intricacies and subtleties of academic analyses of those scenarios would have been poor sources for his screenplays. Popular histories were much more fitting for his purpose, providing him with straightforward anecdotes, already stripped of complications, which could easily be plugged in to meet the narrative demands of his scripts. If many other recent television screenplays have used popular fiction as sources, Hirst was on to something when he sought story lines, even at one or two removes, from "true" history. As many shows with contemporary settings exemplify, stories "ripped from the headlines," or from the judicial and quasi-judicial sources of the past, are compelling. And they are compelling both because viewers have pleasure in knowing what they see on the screen "really happened," and because the stories have *already* been given a narrative form in the original historical records. The archives of medieval and Tudor England are filled with ripping yarns, because those who drew up the documents in the first place had to convince the original intended reader (a judge, a government official) to take a particular action.[67] A riveting tale was the best way to do this. Lawyers and bureaucrats of the medieval and early modern past knew, just as 21st-century screenwriters know, how to tell a good story.

In *The Tudors*, Hirst settled on a formula that reflects and develops recent popular cultural uses of the reign of Henry VIII. The differences between the Tudorism exemplified by *The Tudors* and the television medievalisms of similar recent series are often subtle, as many of the same themes and filmic techniques are on display. *The White Queen*, for instance, which aired in 2013 on the BBC in the UK and on Starz in the U.S., was marketed to some extent as a prequel to *The Tudors*.[68] Based on the 15th-century Wars of the Roses, it has a similar look and features some of the same actors; its early episodes focus on the same narrative arc as the first seasons of *The Tudors* (beautiful commoner defies handsome king's demand that she come to his bed as his mistress, to be rewarded for her resistance with marriage and a crown). Yet although the two series, set a mere half-century apart, have much in common with one another and appeal to the same audience, *The White Queen* is situated much more firmly outside of modernity than is *The Tudors*, which repeatedly emphasizes itself as a genealogy of the present. Thus *The Tudors* includes a scene, set in 1535, where the "new" invention of the printing press is unveiled, accompanied by a portentous announcement that "it will change the world"[69]—even though the printing press was invented about 1450 and had come to England in the 1470s. The new world-changing invention would, in fact, have fit perfectly in a chronological sense into *The White Queen*, set between the 1460s and the 1480s, but such forward-looking technology would disrupt the medieval feel of that series.

As the printing press exemplifies, the popular distinctions between what feels "medieval" and what feels "Tudor" are sometimes at odds with historical chronology: historically, for instance, accusations of witchcraft were far more prevalent in the 16th century than in the 15th, and yet in the popular imagination it is the medieval period, rather than the reign of Henry VIII or Elizabeth, that connotes magic and witchcraft.[70] Thus story lines in *The White Queen* are suffused with magic and the supernatural, while in *The Tudors* all plot motors are strictly anchored in this world. Elizabeth Woodville (the titular character in *The White Queen*) and Anne Boleyn were both accused of using necromancy to ensnare their royal husbands; in *The White Queen*, Elizabeth Woodville really is a witch who literally enchants Edward IV, while the viewer of *The Tudors* knows that the charges against Anne Boleyn are without basis.[71] Whatever Anne's faults, they are modern faults of over-ambition and greed, not primitive manipulation of the supernatural. The characters in *The Tudors* act in ways recognizable to viewers; throughout *The Tudors*, the 16th century is signaled as the beginning of "us." This is a theme of Tudorist popular culture: with religious change, New World explorations, and a growing sense of national identity, 16th-century England stands in the popular imagination as a liminal period at the entryway to modernity. Plotlines involving the Reformation, New World voyages, and new technology run through Tudorist works, including *The Tudors*, often in ways that express ambiguity about the nature of those changes.[72]

Although in subtle ways Tudorist popular culture differs from its medieval counterpart—especially in its liminal position at the dawn of the secular and technological modern—*The Tudors* has a great deal in common with the medieval costume dramas presented on British and North American television in recent years. These series feature sumptuous production values and big-name actors, but scripts that lack nuance. Hirst and other creators—whether they set their stories in the Tudor period or the medieval, or indeed in other far-off places or long-ago times—use the distant past mostly as a place to project fantasies about simple categories and unambiguous choices. Men are men, women are women, and the brave and honorable are clearly demarcated from the villainous. Moral problems are presented in black and white; humor and irony are absent.[73] If technology, rationality, and even sexual pleasure are figured as modern, so also are troublesome complexity, ambiguity, and uncertainty. As Hirst puts it,

> I'm not very good at writing about contemporary society.... William James said that for a baby, the world is a buzzing, booming chaos and it's like that for me; I can't make much sense of it. I feel more comfortable with history.[74]

Hirst may situate *The Tudors* on the threshold of modernity, but he explicitly chooses not to enter fully into the unstable and insecure place that is the modern world.

Notes

1. Acknowledgments: I would like to thank Kit French, Eric Reiter, Tim Stretton, Karolyn Kinane, and Meriem Pagès for their suggestions and comments on this essay, and the history students at Concordia who invited me to present an early version of it to them.

Tatiana C. String, "Myth and Memory in Representations of Henry VIII, 1509–2009," in *Tudorism: Historical Imagination and the Appropriation of the Sixteenth Century*, ed. Tatiana C. String and Marcus Bull (Oxford: Oxford University Press for the British Academy, 2011), esp. 201–206. The original Whitehall mural portrait, painted in 1537, was destroyed by a fire, but was frequently copied in the years that followed its creation. The Walker Gallery in Liverpool holds an early copy: "The Walker Art Gallery's Henry VIII," www.liverpoolmuseums.org.uk/walker/exhibitions/henry/walker.aspx.

2. Alexander Korda, *The Private Life of Henry VIII* (United Artists, 1933). For other influential screen depictions of Henry, see Thomas S. Freeman, "A Tyrant for All Seasons: Henry VIII on Film," in *Tudors and Stuarts on Film: Historical Perspectives*, ed. Susan Doran and Thomas S. Freeman (Houndmills: Palgrave Macmillan, 2009), 30–45.

3. "The Buttery King Henry VIII Feast," http://www.infoniagara.com/dining/buttery/buttery_king_henryVIII.aspx; "Snickers Greensleeves Commercial," http://youtu.be/4yuouH4U_6c.

4. Matt Groening, "Margical Mystery Tour," *The Simpsons* (Fox, February 8, 2004). The Henry VIII story makes up one of three mini-stories told in the episode.

5. Michael Hirst, *The Tudors* (Showtime/CBC/BBC2/TV3 Ireland, 2007–2010). According to an interview Susan Bordo conducted with Natalie Dormer, who played Anne Boleyn, the producers also wanted Anne to be blonde, and only agreed that she should be dark (as the real Anne was) after Dormer lobbied them. Susan Bordo, *The Creation of Anne Boleyn: A New Look at England's Most Notorious Queen* (Boston: Houghton Mifflin Harcourt, 2013), 204–5.

6. "Michael Hirst," *United Agents: The Literary and Talent Agency*, http://unitedagents.co.uk/michael-hirst#profile-4.

7. "The Tudors," Metacriticwww, www.metacritic.com/tv/the-tudors; for fan response, see *The Tudors Wiki*, www.thetudorswiki.com.

8. https://www.facebook.com/TheTudors.

9. For a good summary of Henry's biography and for further references, see E. W. Ives, "Henry VIII (1491–1547)," ed. H. C. G Matthew and Brian Harrison, *Oxford Dictionary of National Biog-*

raphy (Oxford: Oxford University Press, 2010); and Kevin Sharpe, *Selling the Tudor Monarchy: Authority and Image in Sixteenth-Century England* (New Haven: Yale University Press, 2009), 159–61, 171–72.

10. Bordo, *Creation of Anne Boleyn*, 206.

11. Mike Hale, "Blood on Their Hands, and Sex on Their Minds," *The New York Times*, July 22, 2010, sec. Television, http://tv.nytimes.com/2010/07/23/arts/television/23pillars.html.

12. See especially Tatiana C. String and Marcus Bull, "Introduction," in *Tudorism,* ed. String and Bull, 1–12, and the other essays in the volume; see also Mark Rankin, Christopher Highley, and John N. King, eds., *Henry VIII and His Afterlives: Literature, Politics, and Art* (Cambridge: Cambridge University Press, 2009); Bordo, *Creation of Anne Boleyn.*

13. Gary R. Edgerton, "Television as Historian: A Different Kind of History Altogether," in *Television Histories: Shaping Collective Memory in the Media Age*, ed. Gary R. Edgerton and Peter C. Rollins (Lexington: University Press of Kentucky, 2001), 6–7; Erin Bell, "Televising History: The Past(s) on the Small Screen," *European Journal of Cultural Studies* 10, no. 1 (2007): 5. For the argument that fidelity does matter, see Thomas S. Freeman, "It's Only a Movie," in *Tudors and Stuarts on Film: Historical Perspectives*, ed. Susan Doran and Thomas S. Freeman (Houndmills: Palgrave Macmillan, 2009), 1–28; David Puttnam, "Has Hollywood Stolen Our History?," in *History and the Media*, ed. David Cannadine (Houndmills: Palgrave Macmillan, 2004), 160–66.

14. Jerome De Groot, "Slashing History: The Tudors," in *Tudorism,* ed. String and Bull, 243–60 at 244, 250.

15. String, "Myth and Memory," 220.

16. De Groot, "Slashing History," 243–44, 247–50, 259–60.

17. Richard Burt, "Getting Schmedieval: Of Manuscript and Film Prologues, Paratexts, and Parodies," *Exemplaria* 19, no. 2 (2007): 217–19.

18. See, for instance, television critic Charlie Brooker's remarks: the show "may or may not be accurate," but regardless, Jonathan Rhys Meyers's version of Henry is "not a fascinating villain, or even just a flawed human being, but a twat. I'm giving him two more episodes to show some redeeming qualities. Or even just mildly interesting ones. And if he can't manage that, he can sod off back to Tudorland. Or wherever it was King Henry came from." "Charlie Brooker's Screen Burn," *The Guardian*, October 13, 2007, sec. Media, http://www.guardian.co.uk/media/2007/oct/13/comment.tvandradioarts.

19. Freeman, "Tyrant for All Seasons," 5–7.

20. Burt, "Getting Schmedieval," 217–18; Tison Pugh and Angela Jane Weisl, *Medievalisms: Making the Past in the Present* (London: Routledge, 2012), 86–87.

21. In 2008 when the feature was made, Stammers was a doctoral student in 19th-century French history at Cambridge. http://www.hist.cam.ac.uk/directory/tes27@cam.ac.uk.

22. Chris Curtis, "Michael Hirst, The Tudors," 21 May 2009, http://www.broadcastnow.co.uk/michael-hirst-the-tudors/5001701.article. Hirst completes this thought when, in the series' final episode, he has Holbein paint the portrait as Henry is on his deathbed in 1547.

23. Curtis, "Michael Hirst."

24. "Strange but True?" *The Tudors Wiki*, n.d., http://www.thetudorswiki.com/page/Strange+but+True+%3F.

25. Quoted in Burt, "Getting Schmedieval," 236; for other examples, see Felix Schröder, *The Making of The Pillars of the Earth*, DVD (Tandem Communications, 2010); Steven Jack, *History vs. Hollywood: Kingdom of Heaven*, Documentary, History Channel, 2005.

26. Pollard operates through the firm Visual Artefact, www.visualartefact.com.

27. "Jenny Alexander (V)," www.imdb.com/name/nm4129541.

28. Brian Winston, "Combatting 'A Message without a Code': Writing the 'History' Documentary," in *Televising History: Mediating the Past in Postwar Europe*, ed. Erin Bell and Ann Gray (New York: Palgrave Macmillan, 2010), 45–46.

29. Freeman, "It's Only a Movie," 5–7.

30. Curtis, "Michael Hirst."

31. Lina Das, "Lie Back and Think of Olde England! Is This TV's Sexiest Historical Romp?" *Mail Online*, September 7, 2007, http://www.dailymail.co.uk/tvshowbiz/reviews/article-480475. Ridley Scott makes a very similar remark in Jack, *History vs. Hollywood: Kingdom of Heaven*.

32. More charitably, Tom Betteridge has suggested that "*The Tudors* is an exemplary piece of

postmodern history. It desires authenticity, makes it a fetish, while at the same time denying its possibility." "Henry VIII and Popular Culture," in *Henry VIII and His Afterlives*, ed. Rankin, Highley, and King, 215.

33. Das, "Lie Back."

34. Tom Betteridge comments on how closely the series follows reported speech in its account of the fall of Anne Boleyn. "Henry VIII and Popular Culture," 214–15.

35. For the original case, see Shannon McSheffrey, "The Slaying of Sir William Pennington: Legal Narrative and the Late Medieval English Archive," *Florilegium* 28 (2011): 169–203. Hirst's source is likely the brief account in Alison Weir, *Henry VIII: The King and His Court* (New York: Random House, 2001), 314.

36. Kew, The National Archives [TNA], SP 1/124, fol. 204r; SP 1/125, fols. 40r-43v; SP 1/127, fol. 201r. I found reference to the manuscript records through the published calendar (summary) of the State Papers known as *The Letters and Papers* (J. S. Brewer, James Gairdner, and R. H. Brodie, eds., *Letters and Papers, Foreign and Domestic, of the Reign of Henry VIII* [London: Longman, Green, Longman & Roberts, 1862], 12/2:243, 273, 491); as below, I later discovered that G. R. Elton had discussed the records in *Policy and Police: The Enforcement of the Reformation in the Age of Thomas Cromwell* (Cambridge: Cambridge University Press, 1972), 10–11.

37. J. H. Baker, *The Oxford History of the Laws of England, Volume VI, 1485–1558* (Oxford: Oxford University Press, 2003), 544–51.

38. TNA, SP 1/124, fol. 204r.

39. TNA, SP 1/125, fols. 40r-43v.

40. TNA, SP 1/127, fol. 201r.

41. He was still acting in this capacity in 1544: TNA, KB 27/1131, rex m. 6.

42. Hirst, *The Tudors*, S2E5.

43. Elton, *Policy and Police*, 10–11.

44. "Strange but True?" Another popular biographer, Carrolly Erickson, also tells the story; Carolly Erickson, *Great Harry* (New York: St. Martin's Griffin, 1980), 253. Weir's version seems more likely to be the source for Hirst's scene.

45. Weir, *Henry VIII*, 377.

46. *Letters and Papers*, 12/2:243, 273, 491.

47. Graeme Rimer, Thom Richardson, and J. P. D. Cooper, eds., *Henry VIII: Arms and the Man, 1509–2009* (Leeds: Royal Armouries, 2009), 95. I have calculated the size from the waist measurement.

48. Gabrielle Donnelly, "Question Time with Jonathan Rhys Meyers," *Mail Online*, May 16, 2009, sec. Femail, http://www.dailymail.co.uk/femail/article-1180855.

49. Weir's notes refer only to the general work in which a source can be found, so that the reference for the passage quoted above simply reads (with no volume or page number) "L&P"—for *Letters and Papers of Henry VIII*, a collection that has about thirty-seven volumes, some of them over 1000 pages, a less-than-precise footnoting system. Weir, *Henry VIII*, 598 n. 13.

50. For a discussion of the influence of earlier popular histories on films such as Korda's *Private Life* and *A Man for All Seasons*, see Freeman, "Tyrant for All Seasons," 34–35.

51. For a survey of historians' views of Henry's sex life, see Greg Walker, "'A Great Guy with His Chopper'? The Sex Life of Henry VIII on Screen and in the Flesh," in *Tudorism*, ed. String and Bull, 236–37.

52. Elton, *Policy and Police*, 9–12; Walker, "Great Guy," 235–40.

53. Ives, "Henry VIII."

54. Sharpe, *Selling the Tudor Monarchy*, 137.

55. Tatiana C. String, "Projecting Masculinity: Henry VIII's Codpiece," in *Henry VIII and His Afterlives*, ed. Rankin, Highley, and King, 148–51.

56. See "Still of Jonathan Rhys Meyers in *The Tudors*," *Internet Movie Database*, http://www.imdb.com/media/rm2409073408/nm0001667.

57. There are two major gay story lines: Sir William Compton, one of the king's retinue, has a sweet love affair with the musician Thomas Tallis before he (Compton) dies an untimely death from the sweating sickness; George Boleyn, Anne's brother, has a more ambivalent relationship with another musician, Mark Smeaton. The series does not ever mention England's first statute (sponsored and heavily favored by the king) making "buggery" a crime, one that was passed in

1533; however, in both gay story lines those involved are careful to be discreet as it is assumed that the relationships must be concealed. On the statute, see Stanford E. Lehmberg, *The Reformation Parliament 1529–1536* (Cambridge: Cambridge University Press, 1970), 185.

58. Alain Boureau, *The Lord's First Night: The Myth of the Droit de Cuissage* (Chicago: University of Chicago Press, 1998). As Boureau notes, the "droit de cuissage" (the French form of the term) was frequently invoked in media discussion of sexual harassment in France from the 1980s onwards.

59. The screenwriter for *Braveheart* seems to be responsible for inventing the term "prima nocta," which is a nonsensical rendering of the Latin phrase *jus primae noctis*. It is now perhaps the most common form of the term in English. This illustrates the influence of the film in spreading the concept, although at the same time the myth-breaking propensities of the internet have served to create more pages debunking the "right" as defining it, as a google search of the phrase "prima nocta" shows.

60. James Kent, Jamie Payne, and Colin Teague, *The White Queen* (USA/UK: BBC/Starz, 2013).

61. Sergio Mimica-Gezzan, *The Pillars of the Earth*, DVD, Pillars of the Earth (Tandem Communications, 2010), episode 1.

62. Michael Hirst, *Vikings*, History Channel, 2013. There are rapes, attempted rapes, and other violence engendered by sex in episodes 1, 2, and 4.

63. Michael Caton-Jones, *World Without End*, DVD (Canada, Germany, U.K.: Tandem Communications, 2012). There are rapes and near-rapes in episodes 1, 2, 3, 4, 6, and 7 (sometimes more than one per episode).

64. One involves George Boleyn, Anne's gay brother, who anally rapes the woman whom he has been forced to marry on their wedding night (S3E6); the other follows the standard aristocratic rape script as Thomas Culpepper, who will become Queen Catherine Howard's lover, comes upon the wife of a park keeper, and finding her alone, forces himself on her simply because he can (S4E1).

65. Ramona Wray, "Henry's Desperate Housewives: The Tudors, the Politics of Historiography and the Beautiful Body of Jonathan Rhys Meyers," in *The English Renaissance in Popular Culture: An Age for All Time*, ed. Gregory M. Colón Semenza (New York: Palgrave Macmillan, 2010), 30–31.

66. One historical figure whom Hirst probably should have put in the "no-sex" category, but made instead into a "mindless tart" (as Susan Bordo puts it) was Marguerite of Navarre, the sister of the French king Francis I. Marguerite was an intellectual known for her piety and virtue, a patron of great significance for the development of the French Renaissance. In *The Tudors*, she is a busty minx whose brief appearance (S1E4) involves her enthusiastically jumping into Henry's bed. Bordo, *Creation of Anne Boleyn*, 206–7.

67. See Paul Gewirtz, "Narrative and Rhetoric in the Law," in *Law's Stories: Narrative and Rhetoric in the Law*, ed. Peter Brooks and Paul Gewirtz (New Haven: Yale University Press, 1996), 2–13.

68. See Glenda Cooper, "Will the Plantagenets Now Topple the Tudors?," *Telegraph.co.uk*, May 31, 2013, http://www.telegraph.co.uk/culture/tvandradio/bbc/10091651/Will-the-Plantagenets-now-topple-the-Tudors.html; Jolie Lash, "The White Queen Q&A: Max Irons Talks Taking On The War Of The Roses," *Access Hollywood*, accessed January 17, 2014, http://www.accesshollywood.com/the-white-queen-qanda-max-irons-talks-taking-on-the-war-of-the-roses_article_83944.

69. *The Tudors*, S2E6.

70. Anita Obermeier, "Witches and the Myth of the Medieval Burning Times," in *Misconceptions About the Middle Ages*, ed. Stephen J. Harris and Bryon Lee Grigsby (New York: Routledge, 2008), 226–237.

71. Elizabeth's spells feature in virtually every episode of *The White Queen*; the accusations against Anne Boleyn fall in Season 2, episodes 8 and 9, of *The Tudors*. Historians disagree whether Anne Boleyn was in fact accused of sorcery; see G. W. Bernard, "The Fall of Anne Boleyn," *The English Historical Review* 106, no. 420 (July 1, 1991): 584–610; Retha M. Warnicke, "The Fall of Anne Boleyn Revisited," *The English Historical Review* 108, no. 428 (July 1, 1993): 653–665. The accusations against Elizabeth Woodville were made in Parliament in 1484: Chris Given-Wilson, ed., *The Parliament Rolls of Medieval England* (London: British History Online, 2010), http://www.british-history.ac.uk/report.aspx?compid=116561 (6:240).

72. See especially String and Bull, "Introduction."
73. Freeman, "It's Only a Movie," 18–19.
74. Curtis, "Michael Hirst." (Hirst misquotes James.)

Bibliography

Archival Records

Kew, The National Archives
Coram Rege Rolls, Court of King's Bench (KB 27)
State Papers (SP 1)

Film and Television Sources

Caton-Jones, Michael. *World Without End.* DVD. Canada, Germany, U.K.: Tandem Communications, 2012.
Groening, Matt. "Margical Mystery Tour." *The Simpsons.* Fox, February 8, 2004.
Hirst, Michael. *The Tudors.* Showtime/CBC/BBC2/TV3 Ireland, 2007–2010.
_____. *Vikings.* History Channel, 2013.
Jack, Steven. *History vs. Hollywood: Kingdom of Heaven.* Documentary, History Channel, 2005.
Kent, James, Jamie Payne, and Colin Teague. *The White Queen.* BBC/Starz, 2013.
Korda, Alexander. *The Private Life of Henry VIII.* United Artists, 1933.
Mimica-Gezzan, Sergio. *The Pillars of the Earth.* DVD. Pillars of the Earth. Tandem Communications, 2010.
Schröder, Felix. *The Making of The Pillars of the Earth.* DVD. Tandem Communications, 2010.

Print and Internet Sources

Baker, J. H. "The English Law of Sanctuary." *Ecclesiastical Law Journal* 2 (1990): 8–13.
_____. *The Oxford History of the Laws of England, Volume VI, 1485–1558.* Oxford: Oxford University Press, 2003.
Bell, Erin. "Televising History: The Past(s) on the Small Screen." *European Journal of Cultural Studies* 10, no. 1 (2007): 5–12.
Bernard, G. W. "The Fall of Anne Boleyn." *The English Historical Review* 106, no. 420 (July 1, 1991): 584–610.
Betteridge, Tom. "Henry VIII and Popular Culture." In *Henry VIII and His Afterlives: Literature, Politics, and Art,* edited by Mark Rankin, Christopher Highley, and John N. King, 208–22. Cambridge: Cambridge University Press, 2009.
Bordo, Susan. *The Creation of Anne Boleyn: A New Look at England's Most Notorious Queen.* Boston: Houghton Mifflin Harcourt, 2013.
Boureau, Alain. *The Lord's First Night: The Myth of the Droit de Cuissage.* Chicago: University of Chicago Press, 1998.
Brewer, J. S., James Gairdner, and R. H. Brodie, eds. *Letters and Papers, Foreign and Domestic, of the Reign of Henry VIII.* 21 vols. in 35 parts. London: Longman, Green, Longman & Roberts, 1862.
Brooker, Charlie. "Charlie Brooker's Screen Burn." *The Guardian,* 13 Oct. 2007, sec. Media. http://www.guardian.co.uk/media/2007/oct/13/comment.tvandradioarts.
Burt, Richard. "Getting Schmedieval: Of Manuscript and Film Prologues, Paratexts, and Parodies." *Exemplaria* 19, no. 2 (2007): 217–42.
Cooper, Glenda. "Will the Plantagenets Now Topple the Tudors?" *Telegraph.co.uk,* May 31, 2013. http://www.telegraph.co.uk/culture/tvandradio/bbc/10091651/Will-the-Plantagenets-now-topple-the-Tudors.html.
Curtis, Chris. "Michael Hirst, The Tudors." May 21, 2009. http://www.broadcastnow.co.uk/michael-hirst-the-tudors/5001701.article.
Das, Lina. "Lie Back and Think of Olde England! Is This TV's Sexiest Historical Romp?" *Mail Online,* September 7, 2007. http://www.dailymail.co.uk/tvshowbiz/reviews/article-480475/Lie-think-Olde-England-Is-TVs-sexiest-historical-romp.html.
De Groot, Jerome. "Slashing History: The Tudors." In *Tudorism: Historical Imagination and the*

Appropriation of the Sixteenth Century, edited by Tatiana C. String and Marcus Bull, 243–60. Proceedings of the British Academy 170. Oxford: Oxford University Press for the British Academy, 2011.

Donnelly, Gabrielle. "Question Time with Jonathan Rhys Meyers." *Mail Online*, May 16, 2009, sec. Femail. http://www.dailymail.co.uk/femail/article-1180855/Question-time-Jonathan-Rhys-Meyers.html.

Edgerton, Gary R. "Television as Historian: A Different Kind of History Altogether." In *Television Histories: Shaping Collective Memory in the Media Age*, edited by Gary R. Edgerton and Peter C. Rollins, 1–16. Lexington: University Press of Kentucky, 2001.

Elton, G. R. *Policy and Police: The Enforcement of the Reformation in the Age of Thomas Cromwell*. Cambridge: Cambridge University Press, 1972.

Erickson, Carolly. *Great Harry*. New York: St. Martin's Griffin, 1980.

Facebook, "The Tudors TV Show." Accessed 14 May 2013, https://www.facebook.com/TheTudors.

Freeman, Thomas S. "A Tyrant for All Seasons: Henry VIII on Film." In *Tudors and Stuarts on Film: Historical Perspectives*, edited by Susan Doran and Thomas S. Freeman, 30–45. Houndmills: Palgrave Macmillan, 2009.

_____. "It's Only a Movie." In *Tudors and Stuarts on Film: Historical Perspectives*, edited by Susan Doran and Thomas S. Freeman, 1–28. Houndmills: Palgrave Macmillan, 2009.

Gewirtz, Paul. "Narrative and Rhetoric in the Law." In *Law's Stories: Narrative and Rhetoric in the Law*, edited by Peter Brooks and Paul Gewirtz, 2–13. New Haven: Yale University Press, 1996.

Given-Wilson, Chris, ed. *The Parliament Rolls of Medieval England*. London: British History Online, 2010.

Hale, Mike. "Blood on Their Hands, and Sex on Their Minds." *The New York Times*, July 22, 2010, sec. Television. http://tv.nytimes.com/2010/07/23/arts/television/23pillars.html.

Info Niagara. "The Buttery King Henry VIII Feast." Accessed 18 Aug. 2014, http://www.infoniagara.com/dining/buttery/buttery_king_henryVIII.aspx.

Internet Movie Database. www.imdb.com.

Ives, E. W. "Henry VIII (1491–1547)." Edited by H. C. G Matthew and Brian Harrison. *Oxford Dictionary of National Biography*. Oxford: Oxford University Press, 2010.

Lash, Jolie. "The White Queen Q&A: Max Irons Talks Taking on the War of the Roses." *Access Hollywood*. Accessed January 17, 2014. http://www.accesshollywood.com/the-white-queen-qanda-max-irons-talks-taking-on-the-war-of-the-roses_article_83944.

Lehmberg, Stanford E. *The Reformation Parliament 1529–1536*. Cambridge: Cambridge University Press, 1970.

Mars Corporation. "Snickers Greensleeves Commercial." Accessed 18 Aug. 2014, http://youtube/4yuouH4U_6c.

McSheffrey, Shannon. "The Slaying of Sir William Pennington: Legal Narrative and the Late Medieval English Archive." *Florilegium* 28 (2011): 169–203.

Metacritic.com. "The Tudors." Accessed 14 May 2013, http://www.metacritic.com/tv/the-tudors.

Obermeier, Anita. "Witches and the Myth of the Medieval Burning Times." In *Misconceptions about the Middle Ages*, edited by Stephen J. Harris and Bryon Lee Grigsby, 226–37. New York: Routledge, 2008.

Pugh, Tison, and Angela Jane Weisl. *Medievalisms: Making the Past in the Present*. London: Routledge, 2012.

Puttnam, David. "Has Hollywood Stolen Our History?" In *History and the Media*, edited by David Cannadine, 160–66. Houndmills: Palgrave Macmillan, 2004.

Rankin, Mark, Christopher Highley, and John N. King, eds. *Henry VIII and His Afterlives: Literature, Politics, and Art*. Cambridge: Cambridge University Press, 2009.

Rimer, Graeme, Thom Richardson, and J. P. D. Cooper, eds. *Henry VIII: Arms and the Man, 1509–2009*. Leeds: Royal Armouries, 2009.

Sharpe, Kevin. *Selling the Tudor Monarchy: Authority and Image in Sixteenth-Century England*. New Haven: Yale University Press, 2009.

String, Tatiana C. "Myth and Memory in Representations of Henry VIII, 1509–2009." In *Tudorism: Historical Imagination and the Appropriation of the Sixteenth Century*, edited by Tatiana C. String and Marcus Bull, 201–21. Proceedings of the British Academy 170. Oxford: Oxford University Press for the British Academy, 2011.

_____. "Projecting Masculinity: Henry VIII's Codpiece." In *Henry VIII and His Afterlives: Literature, Politics, and Art*, edited by Mark Rankin, Christopher Highley, and John N. King, 143–59. Cambridge: Cambridge University Press, 2009.

String, Tatiana C., and Marcus Bull. "Introduction." In *Tudorism: Historical Imagination and the Appropriation of the Sixteenth Century*, edited by Tatiana C. String and Marcus Bull, 1–12. Proceedings of the British Academy 170. Oxford: Oxford University Press for the British Academy, 2011.

The Tudors Wiki. Accessed 14 May 2013, http://www.thetudorswiki.com.

United Agents: The Literary and Talent Agency. "Michael Hirst." Accessed 18 Aug. 2014, http://unitedagents.co.uk/michael-hirst#profile-4.

Walker Art Gallery. "The Walker Art Gallery's Henry VIII." www.liverpoolmuseums.org.uk/walker/exhibitions/henry/walker.aspx (n.d.; accessed 18 Aug. 2014).

Walker, Garthine. "Rereading Rape and Sexual Violence in Early Modern England." *Gender and History* 10, no. 1 (1998): 1–25.

Walker, Greg. "'A Great Guy with His Chopper'? The Sex Life of Henry VIII on Screen and in the Flesh." In *Tudorism: Historical Imagination and the Appropriation of the Sixteenth Century*, edited by Tatiana C. String and Marcus Bull, 223–41. Proceedings of the British Academy 170. Oxford: Oxford University Press for the British Academy, 2011.

Warnicke, Retha M. "The Fall of Anne Boleyn Revisited." *The English Historical Review* 108, no. 428 (July 1, 1993): 653–65.

Weir, Alison. *Henry VIII: The King and His Court*. New York: Random House, 2001.

Winston, Brian. "Combatting 'A Message without a Code': Writing the 'History' Documentary." In *Televising History: Mediating the Past in Postwar Europe*, edited by Erin Bell and Ann Gray, 42–58. New York: Palgrave Macmillan, 2010.

Wray, Ramona. "Henry's Desperate Housewives: The Tudors, the Politics of Historiography and the Beautiful Body of Jonathan Rhys Meyers." In *The English Renaissance in Popular Culture: An Age for All Time*, edited by Gregory M. Colón Semenza, 25–42. New York: Palgrave Macmillan, 2010.

Nature and Adventure in *Die Jagd nach dem Schatz der Nibelungen*

Evan Torner

"Do you really believe there's a map on the back of [the Declaration of Independence]?" "No. It's just a plot to make us learn history."—two kids conversing in a museum in the alternate ending to *National Treasure* (2004)

On August 31, 2008, the made-for-TV adventure film *Die Jagd nach dem Schatz der Nibelungen* (*The Charlemagne Code*, hereafter referred to as *JSN*) premiered on the German private television network RTL to much hype and fanfare. With a 4.85 million euro price-tag, the film was an expensive undertaking by genre entertainment company Dreamtool Entertainment, and one that nevertheless delivered returns on investment: beyond World Cup match broadcasts, RTL's ratings had never been higher than that the night the film premiered. *JSN* attracted enough domestic viewers in subsequent broadcasts that, in 2010, it was specially selected to screen in China as well.[1] The film's success was followed up with two modestly successful sequels: *Die Jagd nach der Heiligen Lanze* (*The Hunt for the Holy Lance*, 2010) and *Die Jagd nach dem Bernsteinzimmer* (*The Hunt for the Amber Room*, 2012). Though one could see the film's meteoric success as a private television station coup given an increasingly competitive European Union media market, it is worthwhile to note that the lavish budget of *Die Jagd nach dem Schatz der Nibelungen* was only made possible thanks to generous subsidies from the publicly funded North Rhine–Westphalian (NRW) and Bavarian (FFF Bayern) film foundations. That is to say, German regional and national interests also achieved expression in a for-profit production that, on the surface, appears to be little more than a gimmicky German rip-off of Hollywood adventure films *Raiders of the Lost Ark* (1981), *Tomb Raider* (2001), *National Treasure* (2004) and *The Da Vinci Code* (2006). When asked if they have seen the film, many German television viewers I have interviewed vaguely remember catching part of the production as a rerun but, five years later, most categorize it as a decidedly trivial production: "Popcorn TV,"[2] or "fun, amusing and proper entertainment, something rarer and rarer in the German television landscape."[3] Nevertheless, my essay here poses two fundamentally non-trivial questions, namely: What are the national and transnational stakes of a film like *JSN*? And how does such a film reconfigure the notion of the "medieval" to suit modern needs and fantasies? I posit that, in the last decade, Germany has been able to offer a post-ironic[4] interpretation of its own cultural heritage–vacillating between ironic and sincere modes of representation–and that *JSN* exemplifies this trend by framing the academic search for lost medieval history as a tongue-in-cheek cinematic thrill ride. By the same token, however, the very substance of Central European medieval history rebels against its crass appropriation here, both in the film's plotline itself–as the protagonists are denied the Nibelungen hoard in the end–and in the consideration of the sites visited to unlock the treasure. After all, the Teutoburger Forest, Charlemagne's Cathedral at Aachen, the Cologne Cathedral (Kölner Dom), the isle of Ralswick (Rügen) and Castle Neuschwanstein do *not* share similar historical resonances, unless one lumps them together under a rubric of pre–20th-century medievalist icons. My analysis contextualizes the film in terms of its generic, affective and historical qualities, as well as touches on the thorny epistemological questions about the nature of national history that the film perhaps unintentionally raises.

Genre Codes of Charlemagne

The outlandish plot of *JSN* already sounds familiar before one sees the film. Archaeologist Eik Meiers (Benjamin Sadler) is on the trail of the legendary treasure hoard from the 13th-century Middle High German poem *Das Nibelungenlied*, a treasure which Charlemagne supposedly found in 777 AD and then re-hid during his late imperial reign in order to protect his subjects from internal discord. "At [Charlemagne's] request," Eik's wife Maria (Milena Dreißig) reads aloud to her daughter Kriemhild (nicknamed "Krimi,"[5] played by Liv Lisa Fries) at the film's opening, "his advisors left behind clues to find the treasure, so that a wiser man in a wiser generation may find it. ... And thus the treasure of the Nibelungen once again descended into the river of history." Just moments after Maria reads these words, however, Eik loses her forever–and presumably fellow archaeologist

André Cabanon (Stephan Kampwirth) as well–to a collapsing cliff on the isle of Ralswick while retrieving the ancient Carolingian amulet called "the Path of Faith." Eik's wife's death causes him to give up archaeology. Eight years later, Eik is brought back to the case by way of his plucky old partner Justus (Fabian Busch) and the by-the-book professional archaeologist Katharina Berthold (Bettina Zimmermann), who are investigating a stolen engraving from the roof of the now-excavated basilica at Ralswick. Their renewed search puts them in the path of the sickly millionaire Heinrich Brenner (New German Cinema's Hark Bohm) and his hireling André, who survived the cliff disaster after all. Brenner seeks the fabled blood of the dragon Siegfried slew to restore his health and give him immortality. The rival archaeology teams find their way to the lost catacombs under the Cologne Cathedral, where they retrieve the "Kaiser's Destiny" gauntlet, but lose Eik's research diary in the process. Nevertheless, the search continues in the Teutoburg Forest, where the Germanic tribes defeated the Romans in 9 AD, specifically at the Externsteine, those free-standing stones that constitute one of the forest's most popular monuments. After finding the next artifact (the "Eye of God" orb) in one of the stones, André confronts Eik and reveals his true culpability in Maria's death. Eik, Katharina and Justus then lose their artifact collection to him in an armed struggle. Charlemagne's Imperial Cathedral (Kaiserdom) in Aachen is their next tourist attraction/destination. Eik seduces an old flame to gain access to the cathedral's treasure room and the Bust of Charlemagne, which he steals to obtain the "Crown of the Empire," a piece of Charlemagne's skull. Brenner fires André due to his erratic behavior and replaces him with Eik, who quickly finds himself rappelling from the Marienbrücke near the Castle Neuschwanstein waterfall with now love-interest Katharina. There they find the Hall of the World, springing and surviving a room-flooding trap. A ray of sunlight then miraculously points to a spot on a perfectly articulated sculpture of the Zugspitze, the highest mountain in Germany and the Nibelungen hoard's resting place. Once the characters discover the actual treasure in a hidden ice cavern, André ambushes the group, shoots Brenner and tries to walk off with the treasure, only to fall through the ice to his doom. The film ends with Justus donning the fabled invisibility cap (Tarnhelm) he took as a souvenir. He realizes that it doesn't make him invisible, but then receives a call to begin their next archaeological adventure: the quest for the Holy Lance.

The details of the plotline above remain strikingly similar to the film's Hollywood counterparts *Raiders of the Lost Ark* and *National Treasure*, so much so that one might see *JSN* as a direct transposition of those movies into the German context. Like *Raiders of the Lost Ark*, *JSN* features a hot-blooded male archaeologist (Indiana Jones/Eik) and his feisty female compatriot (Marion/Katharina) on the trail of a MacGuffin treasure of legend (the Ark of the Covenant/the Nibelungen Hoard) which was hidden away by an ancient conspiracy (Moses' successors/Charlemagne's advisors) that made it only findable by means of shining sunlight through a specially designed lens (the Staff of Ra/the Path of Faith) on a specially designed three-dimensional model (of Alexandria/of the Alps). Like *Raiders*, *JSN* has an evil, sell-out archaeologist as an antagonist (René Belloq/André) who is willing to go to any lengths to obtain the treasure for himself, and is in turn killed more or less by the treasure itself (the "Baroque Wonder" of *Raiders*/Siegfried's sword causing the ice to crack). Similar to *Raiders*, the ancient world is seen as a site of magnificent treasure, and knowledge thereof as the key to unlocking it.

But where *Raiders* offers a genealogical understanding of blockbuster adventure film tropes, *National Treasure* appears to be *JSN*'s direct progenitor with regard to the film's overall structure. Like *National Treasure*, *JSN* starts with an oral tale of the lost treasure that serves as a voice-over for a flashback sequence filled with costumed characters in the distant past busily sacrificing their lives and safeguarding treasure hoards from their greedy contemporaries. Like *National Treasure*, the chief male protagonists (Benjamin Franklin Gates/Eik) are traumatized by their past failures in both academia and their personal lives, such that their treasure-hunting not only redeems them, but also attracts a fellow obsessive scholar as a romantic partner (Abigail Chase/Katharina). Personal narratives of the self achieve resolution through the uncovering of precious metals sculpted and hidden away. Like *National Treasure*, there are a series of artifacts that need to be found at different locations and assembled or interpreted to find the treasure. But *National Treasure* and *JSN* also share a deep investment in fairly normative interpretations of national histories of the United States and Germany respectively. In both films the nation is conceived as a source of treasures to be shared by the collective ... if only a group of brave, unselfish individuals might push their physical and intellectual capacities to the limit to unearth it. Benjamin Franklin and Charlemagne are in this fashion both portrayed as benevolent founding father figures, whose acts of concealing treasures behind puzzles constitute some kind of future investment in a nation-yet-to-come; an extraordinary gesture toward futurity. Such communicative benevolence on both sides solves the problem of the *selfish national subject*: a character who pretends to believe in the nation for the sake of personal gain, of creating a hoard for himself. Medieval and early modern ancestors wisely lock future wealth away from the present, so that their clever modern descendants are not pursuing profit when they seek this wealth. A better justification for neo-liberal enlightened self-interest in a western national context could be scarcely imagined.

Adventure Genre as Emotion System

To say that *JSN* is a shallow derivative of better-produced Hollywood content misses the point that the film, in fact, also takes a page directly from the history of the German adventure film genre.[6] Adventure films have been a part of German film history since the short films of Richard Eichberg and Harry Piel in the 1910s, followed by Fritz Lang's serial *Die Spinnen* (*The Spiders*, 1919/1920) and Joe May's epic *Das indische Grabmal* (*The Indian Tomb*, 1921). Conventional typologies for the adventure genre often divide it into sub-genres, as Brian Taves does: the swashbuckler, pirate, sea, empire and fortune hunter sub-genres all bear the mark of Hollywood.[7] The German sub-genres, however, break down along slightly different lines, namely: the Orientalist adventure (e.g., *Die Abenteuer des Prinzen Achmed*, 1923–26), empire (*Kautschuk*, 1938), mountain (*S.O.S. Eisberg*, 1933), "Indianerfilm" (*Der Schatz im Silbersee*, 1962) and anti-adventure (*Aguirre, Zorn des Gottes*, 1972).[8] Unifying elements of the German iteration include an emphasis on the protagonist's reclaiming of lost honor within bourgeois modernity (as per the 19th-century literary tradition of Karl May and Fritz Steuben), a preoccupation with ostensibly "dying races" in other countries, and the displacement of tropes from German heroic epics (*Heldensagen*) onto exotic scenarios in far-flung locales. Thomas Klein has noted the integral role of sets and costuming as essential elements in bringing an adventurous "labyrinth of the foreign" to life,[9] summarizing the genre in a few pithy nouns: "the fantastic, love, comedy, technology, magic, nostalgia, sentiment, and an orgy for the senses."[10] Gold treasure troves, introduced as "folk legends," often form the MacGuffins that escort the audience into this labyrinth of the foreign. Yet the fortune hunting sub-genre–of which *JSN* is a prime specimen–only gained traction in German film production after the runaway success of Indianerfilm spoof *Der Schuh des Manitu* (*Manitou's Shoe*, 2001) and the campy *Detective Lovelorn und die Rache des Pharaos* (*Detective Lovelorn and the Revenge of the Pharaoh*, 2002), i.e., once the cinema of the Federal Republic of Germany had become well and truly tied to blockbuster imperatives. On a symbolic level, the characters' cracking the code that leads to the treasure parallels the filmmakers' hopes to reap revenue by delivering to the audience what it presumably wants. What *JSN* offers in addition, however, is an iteration of this German genre that actually takes place on German soil, treating Central European medieval heritage as a product to be colonized, nationalized and rendered "exciting" in the limited fashion that the adventure genre permits.

It is thus not enough to dismiss *JSN*'s clichés and overtures as formulaic; rather, we must understand this film *in terms of* the emotional appeals and fantasy propositions about modern German subjects on a treasure hunt by way of their knowledge of the medieval. Fredric Jameson has noted that generic signals actually preserve within them (rather than gloss over) the societal paradoxes they attempt to resolve.[11] Taking a cue from poststructuralism, Jameson sees a careful analysis of popular fantasy propositions as useful because, by his reasoning:

> [What] is wanted is an inventory of the dilemmas of representation, of what in the structure of object or subject alike makes representational accuracy or truth an impossible achievement and an ideological ambition or fantasy as well. We map the contours of globalization negatively, by way of a patient exploration of what cannot be perceived and what cannot be narrated.[12]

Jameson specifically seeks the fictions inherent in the objects and systems that surround us, unraveling the threads that tie together our global system to expose the impossibility of an objective discursive position or "neutral" national conversation. With the use of the "medieval" in film, Tison Pugh and Lynn T. Ramey posit that, in fact, "much can be gleaned from films that 'get it all wrong,' especially since [such] films evince little or no interest in getting the Middle Ages 'right.'"[13] Instead of probing *JSN* for its obvious colportage and its ridiculous plot holes, it appears more reasonable to evaluate the kind of cultural work that such symbols appear to be performing for modern directors and audiences.[14] Much of this cultural work is actually emotion-work. Cinematic and televisual fiction permits us to analyze the mood (affect), personality, motivation and emotion of the figures presented. It allows us the analysis of institutions by means of the dreams these institutions project onto the world, corresponding with Jerome Christensen's argument that films are perfect sites to examine institutional personhood, as their plotlines and imagery allegorize specific studio interests as well as broader-scale social systems.[15] Moving images manipulate audience emotion, but they also betray the interests of the networks interested in said manipulation.

JSN's popularity, as with *Der Schuh des Manitu* before it, prompts a scholarly inquiry into the effectiveness of its specific emotional appeals to a German audience.[16] My argument is that these appeals correspond with a form of German nationalism that is openly mining–rather than interrogating–a shared, trans–European history for material that reinforces the continuity of the German "brand" within the European Union. Instead of addressing what Daniela Berghahn and Claudia Sternberg consider the pivotal cinematic issues of modern-day Europe–e.g., "the long legacy of colonialism, the ongoing process of European integration, the geopolitical repercussions of the collapse of communism, continuing intra–European mobility and the influx of migrants and refugees from across the world"[17]–a film like *JSN* appeals to the audience on four intermingled registers. First and foremost, it appeals to the viewer's sense of spectacle to the extent that it fits into a "cinema of attractions" revitalized in Germany since 2000.[18] Attractive actors serve as the focus for an almost continuous array of tracking shots that otherwise take the viewer through most of the major tourist locales in Germany and their "hidden" underbellies. Multiple action sequences highlight the physically dangerous and visually breathtaking task of tracking down a medieval hoard. The second register appeals to a German sense of history that has curiously repressed any and all references to the Holocaust and the crimes of the Third Reich. Reaching back to Charlemagne and reifying the hoard of the Nibelungen, *JSN* sees German culture through a selective lens that effectively lifts the heavy burden of 20th-century history from the viewer's shoulders. Though the adventure franchise later addresses the Third Reich in its third installment *Die Jagd nach dem Bernsteinzimmer* (2012), its first film explicitly preoccupies itself with a somehow-undisturbed 1,000-year-old archaeological history of Germany that simply omits the intervening years. The third register is the intertextual pleasure that an informed viewer who has seen *Raiders of the Lost Ark* or *National Treasure* experiences in his/her encounter with the film. As pointed out earlier, many of the tropes are similar, such that one can watch the adaptation unfold purely *as* adaptation. The final register constitutes the specific affective and emotional cues that such films tend to use, in homage to the serials that inspired them and in dedication to an audience's uninterrupted "entertainment." It is this register that requires the presence of medieval history as both a mental and physical puzzle to be unlocked, such that the audience feels consistently privy to the "secrets" of the various tourist sites arrayed in the film. Each piece of medieval text, every object, turns into a specific puzzle within the narrative for the characters to solve, and every search for every object succeeds at ramping up the stakes and the majesty of the next hidden wonder.

It should be noted that the formula for transforming the painstaking and cinematically uninteresting work of archaeology into an emotional roller-coaster solidified around *Raiders of the Lost Ark*. As a work of pure industrial filmmaking, *Raiders* is specifically built around an attention economy rationalized for maximum thrills. At an early 1978 script development meeting, filmmakers George Lucas, Steven Spielberg, and Lawrence Kasdan candidly articulated the logic of how the film would generate suspense. George Lucas' contribution was to mathematically plot out how often they could get away with endangering the characters: "One of the main ideas was to have, depending on whether it would be every ten minutes or every twenty minutes, a sort of a cliffhanger situation that we get our hero into."[19] Lucas then turns toward the creation of Indiana Jones, the kind of hero needed to be able to address competently each of these cliffhanger scenarios: "Instead of being a martini drinking cultured kind of sophisticate, he's the sort of intellectual college professor James Bond. He's a superagent."[20] Spielberg later asks whether or not his "main adversaries" will be the Germans, to which Lucas responds, "Yeah, I think they should be. I've been trying to move him around the world a little bit to see if we can't get a little Oriental influence into it just for the fun of it."[21] Spielberg responds with "What we're just doing here, really, is designing a ride at DisneyLand."[22] This conversation proves remarkable, albeit unsurprising, in that it already establishes the formula for adventure films not around the weight of the histories involved, but around a specific emotion system designed to keep viewers on the edge of their seat as much as possible without ever truly endangering their hero.

In light of such facts, *JSN* becomes more than just a genre film-made-for-television, a mere "commercial feature film which, through repetition and variation, tell familiar stories with familiar characters in familiar situations" (as Barry Keith Grant describes it).[23] It is instead, as with most successful genre productions, a highly sophisticated management system of human attention, using the materiality of medieval history as bait. Greg M. Smith sees films as sophisticated apparatuses for "carefully packaging and selling" human emotion.[24] His analysis of *Raiders of the Lost Ark* demonstrates what he perceives to be the interaction between mood and emotion through filmic systems of "mood cues" to prepare the widest possible viewer demographic for bursts of emotion to come. Redundant emotive cues are littered throughout adventure cinema, so that the audience transparently knows what to feel when the filmmaker wants to trigger a specific emotion with an emotion marker. Smith's analysis of the opening sequence of *Raiders*, for example, demonstrates that the shocking, grotesque stone idol that a guide encounters as he leads Indiana Jones through the jungle, in fact, delivers no new story information, but rather primes an audience for further more story-oriented shocks to come.[25] *JSN* has modeled itself off on *Raiders*, and thus adopts a similar set of strategies to address the emotions of a wide German-speaking audience.

How does *JSN* use the materiality of the medieval as mood cues? The film begins with two faded-in-and-out shots of medieval manuscripts lit by flashlight as Maria reads the tale of Charlemagne's treasure aloud to Krimi over a brooding orchestral score. The flickering light continues across multiple shots of actors in medieval clothing, as well as halberds crossing in front of Charlemagne's golden bust in Aachen. The camera tracks across the landscape and a raft of hooded attendants bearing some kind of treasure, which we then learn to be the treasure of the Nibelungen. The treasure is iris-framed in close-up and shown flickering in the torchlight as the title of the film rolls in. These opening 45 seconds mood-cue the medieval world as a place of exotic fantasy and secrets, giving the impression of ninth-century courtiers who deliberately made their history mysterious for the sake of explorers encountering it centuries later. Since the sequence is related as an oral tale, there is a direct transposition of the text from the medieval manuscript (which consists of nothing but illustrations) into the imaginary supposedly playing out in the minds of the characters. Here, the film is demonstrating its ability to enchant the past and invites the viewer along into its fictional alternative history. The discovery of the hoard at the end of the movie placates the viewer's interest, and thus the audience experiences catharsis when the near perfectly-preserved treasure is reburied by a collapsing ice cavern. The treasure has been found, the intention of giving it back to the German people stated, and thus it can be safely returned to obscurity.

Architecture, lighting and physical artifacts also play important roles in cuing the mood. The excavated chapel at Ralswick, for example, is carefully underlit to preserve the shadows on the arches in the underground space. This intimate lighting also cues the viewer for Eik's numerous personal confrontations with his past in the scene: finding his wife's ring and confessing to Katharina about his old archaeological career. Brenner's mansion is similarly underlit, albeit decorated with eerie modern furniture, as André reveals his acquisition of the ancient mural he stole. Underlit areas constitute sites of personal and narrative revelation; here there be secrets and, slowly, they shall be revealed. There is no question then that the narrative resolution of the search for the treasure takes its form as a beam of light, properly reflected around the Hall of the World, and that the Nibelungen hoard in the ice cavern be lit from above. The physical artifacts from the ninth century such as the solidus, the minted coin with a dragon head, also play a role in mood-cuing. On the one hand, they are treated with care and veneration as objects from a hallowed past; on the other hand, they are considered known quantities with which the characters are already extremely familiar and thus are only important insofar as they lead to the treasure. The archaeologists blow into 800-year-old horns, insert gold slivers into stone cracks, and pull off corpses' hands. Their equal parts nonchalance toward and absolute expertise about the artifacts mirror fantasies of dominance over German material history: that the past can be accurately cataloged, felt and even discarded for the sake of posterity. To quote Jacques Derrida, the television-film articulates the fantasy moment when "the *arkhē* appears in the nude, without the archive."[26] In an adventure film, it is important for protagonists to maintain mastery over the objects in their physical environment. Such a disposition opposes the Derridean notion of the archive, which sees it as structured by its contents and the networks of power that surround it, rather than by autonomous human subjects in search of the key to a puzzle. The mood medieval objects and artifacts create in the film is thus always one of intimacy and playfulness, suggesting a ludic attitude opposite to that of L.P. Hartley's overused phrase: "The past is a foreign country: they do things differently there."[27]

In many respects, the past in *JSN* is more like a *colonized* country, silenced and parroted in caricature for the sake of redundantly impacting modern audiences' emotional sensibilities. As I discuss below, the "DisneyLand ride" composed from the original source material has perhaps more serious implications with its emotional propositions than intended. The issue lies in the institutional paradox of the genre adventure film: the emotional cocktail must be delivered with the proper dosage, but also with a veneer of historical "accuracy." Bettina Bildhauer explains:

> Audiences will enjoy a film more if it looks authentic to them, if they can imagine the stories told to have really happened; and especially if they can have this confirmed by experts. However, they will also enjoy a film more if it offers a satisfying artifice with a star adding contemporary or monumental grandeur, beautiful images and exciting narratives–features that draw attention to its filmic nature rather than simulate transparent access to the past.[28]

Thus one can see "realism" as more an affordance of optimized film production processes than an absolute to which the filmmakers had to adhere. Artifice, along with the necessary realistic gloss to suspend disbelief, is the holy grail of both blockbuster cinema and high-budget television productions alike. But unlike blockbuster cinema, as Lisa Parks and Shanti Kumar articulate, television is not so much an entertainment "event" as "it is part of the very social fabric that gives shape to us as individual subjects and imagined communities."[29] That is to say: television as a medium has a fairly high stake in representing other media from past eras as quotidian or already-mastered. Much of the display of medieval scripture, artifacts and constructs signal a past both accessible and easily dismissed as a fairytale. I now turn to the quite-limited historical imaginary of the production itself.

The Persistence of History

Subtending the promise of a blissfully superficial adventure movie is a very serious national-pedagogical project to re-enchant what is seen as a shared German medieval past. There is, indeed, a plot afoot to make German children learn history, and it is a history purged of the messiness of German nationhood. Fortunately, the film helps reveal this project's inherent paradoxes and incoherence. The creation of a specifically German medieval heritage to the exclusion of other linguistic and cultural traditions was a project undertaken most notably by historian Johann Friedrich Böhmer (1795–1863), mythologist Jacob Grimm (1785–1863) and philologist Karl Lachmann (1793–1851). Böhmer and Lachmann rearticulated Germanophone history through Latin and Middle High German texts, while Grimm used oral folktales coupled with library holdings to compile a catalog of what would be considered the German mythological tradition. Böhmer's *Regesta imperii* helped register and collect historical facts about a litany of southern German (Catholic) kings, a counter-move to the increasing dominance of Prussian Protestant discourse in the 19th century. Jacob Grimm's 1835 four-volume *Deutsche Mythologie* was intended as a corrective to the link supposedly severed by Christianity (a Semitic religion) between the German *Volk*, their land, myth and nation.[30]

In the meantime, Lachmann assembled the canon of German poets–particularly Walther von der Vogelweide and Wolfram von Eschenbach—as well as German epic poetry such as the *Nibelungenlied*. Lachmann's painstaking engagement with and publication of these older texts gradually convinced German speakers that they had privileged access to a unified Germanic past, and thus a potentially unified future. Despite Lachmann's overwhelming proof that such access was partial, fragmentary and seen very much through the lens of materials available in the 19th century, philosophers relied on a direct, unified genealogy from the Teutons, Franks and Vandals to the Saxons, Bavarians, and Prussians to argue for German nationhood.[31] Regardless of nationalism's eventual outcomes, including but not limited to the rise of National Socialism, these scholars were doing what has always been done: bargaining for political gains in the present using an essentialized past. The question then arises: what use is *JSN* to which political agendas? How does the essentialized past it portrays serve this agenda?

In fact, there *still* exists a master discourse called "history" that, all claims of postmodernity aside, *still* governs that which we call the "nation" and the phenomenon of "globalization," or those rapid global effects that bypass all national boundaries to affect populations directly. "There is a preoccupation with history when history seems to be increasingly irrelevant to understanding the present," Arif Dirlik writes, responding to the question of the discipline's inherent Eurocentrism. "Worked over by postmodernism, among other things, the past itself seems to be up for grabs, and will say anything we want it to say."[32] The "preoccupation with history" that Dirlik addresses can be found in symptomatic form in popular film and television, specifically the idea that the past is a kind of subterranean natural resource (like oil or diamonds) that can be productively mined for the benefit of the present. Previously under-considered archival texts are treated in the media as "forgotten gems." Archaeological findings present "unearthed treasures" for the visual consumption of a populace eager to find an affective connection to a heritage more capacious than their own. Yet due to the necessities of genre entertainment and fiction that are becoming increasingly game-like,[33] history is often also presented as a *puzzle*, an exercise in abstract symbolic and mathematical logic that an engaged viewer might "solve," as if the fragmented mess of human history might yield increased prosperity for all, if only the brave and clever few might crack the code.[34] On a pragmatic level, the history-as-puzzle trope performs three fundamental functions: to make a film's protagonist appear more intelligent and therefore more "worthy" of finding the treasure, to provide suspense in non-action/dialog sequences, and to (presumably) activate the audience's own creative puzzle-solving capacities to "help" reason out the situation as the protagonist grasps for answers. The archaeological fortune-hunter film genre therefore articulates a specific subjectivity with regard to history: medieval archaeologists and historians are apparently not concerned with social context or the polysemy of language, but rather semantics and etymology as it pertains to the location of treasure. Moreover, *JSN* denies lessons learned from both medieval and 20th-century historical accounts when it moves in the direction of fixed German national boundaries and purges the narrative of any contested notions of Germanness.

It is nigh impossible to divorce any German television or feature film from engagement with Germany's troubling history, such that Thomas Elsaesser once noted that "the German nation is haunted by its cinema screen, and the films are haunted by German history."[35] Discussing recent filmmakers of the last decade, Jaimey Fisher and Brad Prager state that "regardless of their apolitical aims or even, paradoxically, owing to their insistence on aesthetic autonomy, many of today's German films are discussed in political terms."[36] *JSN*'s plot and symbolism have apparently evaded political discussion, a situation that this essay seeks to rectify. We can begin with the political significance of the various sites chosen for Charlemagne's four puzzle components. Ralswick was never occupied by Charlemagne, but rather by the Slavic Rani and later by the Danes. It also represents the one location in the film (besides the Pergamon Museum) that was in the GDR, meaning that this story is primarily about the former West Germany, and its Catholic sites at that. In Cologne, the cathedral was built six full centuries after Charlemagne's death, and its damage in and reconstruction after World War II is deliberately avoided as a topic: the fictional tombs underneath remained apparently untouched by the bombs. The arm reliquary of Charlemagne is "found" there, despite its actual resting place in Aachen. Aachen, Charlemagne's old seat of power, with a far more appropriate vault for the prior artifact, features the theft of the bust of Charlemagne, mirroring Benjamin's theft of the Declaration of Independence in *National Treasure*. The 1349-commissioned gold bust did indeed contain a portion of Charlemagne's skull, but it was commissioned under Aachen's Bohemian rule by Charles IV as a symbolic means of consolidating the Holy Roman Empire, as Charlemagne once did. The Teutoburg Forest was the site of the Hermannschlacht in 9 AD, a battle that presumably (again according to 19th-century German nationalists) kept the Romans at bay. The Externsteine in the Teutoburg Forest were indeed part of Charlemagne's empire, but they also figured prominently in Heinrich Himmler's Ahnenerbe research, and are still a symbol of right-wing neo-pagan movements in Germany. Finally, Neuschwanstein Castle has the least to do with medieval history and the most to do with medievalism. It was constructed in the 19th century as a result of the romanticist revival and the influence of Richard Wagner on Bavarian King Ludwig II. It is also one of the oldest, continuously running castle tourism sites in Germany. Television in this instance offers up a site of virtual tourism. So there you have it: a former GDR resort island, a rebuilt cathedral, another cathedral, the ritual ground for white supremacists, and one of the dream palaces of a romantic Catholic king. And of course, the treasure is buried in the Bavarian Alps. Director Ralf Huettner, who was born, bred, and trained in filmmaking in Munich, betrays his upbringing by symbolically unifying Germany by way of devoutly southern bias.

The underlying significance of these locations is obvious: they are all major tourist destinations in the Federal Republic of Germany, such that the made-for-television film could be seen as an instrumental strategy to re-enchant common tourist destinations for the average German, as the vast tourism industry surrounding the *Da Vinci Code* did for France. The Zugspitze becomes the secret site of the Nibelungen hoard, the Externsteine become the site of Eik and André's epic struggle, and so forth. With the tourism industry as one of the few growth sectors within a substantially post-industrial German economy,[37] German film boards subsidize private-sector adventure films like *JSN* to boost tourism to specific regions, as well as employ German film talent. Films contemporary to *JSN* such as *Inglourious Basterds* (2009) but also specifically "medieval" productions like *Die Päpstin* (*Pope Joan*, 2009) and *Vision—Aus dem Leben von Hildegard von Bingen* (*Vision: From the Life of Hildegard von Bingen*, 2009) were all heavily subsidized to drive tourism and interest to the target regions of Brandenburg, Hessen, and Baden-Württemberg respectively. Thus the fact that Charlemagne's reign extended far beyond German borders—giving most of Europe claim to his national treasure—is ignored due largely to the film's financial underpinnings in Germany. Lack of historical imagination appears to emerge from the film's finance concept. Out of what amounts to censorship of the market, *JSN* delimits Germany (and the sprawling medieval impact of Germanic tribes) so as to build the German tourism brand.

There is another concern, and that is the original universality of the *Nibelungenlied*, and its adaptability to different local topoi.[38] Bernhard Martin has written extensively on the *Nibelungenlied* being "used or more often abused as a means of transmitting and justifying national and/or völkisch patterns of thought."[39] Most notably, Third Reich discourse on the fragmentary epic from the likes of Hans Naumann still pervaded many circles of West German discourse from the 1950s to the present, praising the epic as a national treasure and seeing the hoard itself as a "still hidden, yet-to-be-recovered fund for the nation."[40] In fact, different versions of the *Nibelungenlied* put the location of the hoard in different locales: at the bottom of the Rhine or Siegfried's cellar.[41] In *JSN*, the treasure has moved inexplicably from a subterranean locale to high up in the mountains of Bavaria, its emotional position in the hearts and minds of the viewers unmistakable. Klaus Badelt's musical score turns to choir music when the group finally comes face-to-face with the hoard, and it is this choir that signals some form of collectivity between past and present. One also notices that the original *Nibelungenlied* very deliberately moves the locus of action from West (Xanten, in Nordrhein-Westfalen) to East (Esztergom, near Budapest). In doing so, the text of the epic poem deliberately traverses vast swaths of Europe. The eastern sweep of the epic is replaced by the southern sweep of the film: the group starts at Ralswick and ends at the Zugspitze. On a similar note, Charlemagne was the notorious Christianizer of many Germanic tribes,[42] and the Holy Roman Empire under his rule stretched from Nantes in France to the Spanish Marches, northern Italy, Austria, and parts of Croatia. The emotion-system of the film is dead-set on encountering only objects and cultures from within the confines of present-day, post–1989 Germany: the Frankish king becomes the good German patriarch, his complex legacy subsumed under the more directly nationalized legacy of Emperor Frederick Barbarossa and others who have been folded into the national narrative.

JSN's insular, emotional appeals also adhere to a specific Eurocentric temporality that is ambivalent toward the medieval period. Prasenjit Duara has recently argued that the historiographical practice of periodizing national histories in terms of "ancient," "medieval," and "modern" actually demonizes the "medieval" for the sake of the "modern." Such discourse management serves as a means of "[constructing] a continuous subject by connecting the modern period with the ancient (often by means of a '*renaissance*').... The historian could then bypass the medieval and reject what was unsuitable to modernity as medieval accretions."[43] This tripartite scheme dates back to the mid–14th century and was firmly entrenched in the popular consciousness by historian Christoph Cellarius in 1800.[44] G.W.F. Hegel noted in *The Philosophy of History* that the collapse of Charlemagne's empire after his death more or less ushered in the Middle Ages proper,[45] expressing no small degree of disappointment at that turn of events:

> While the *first* period of the German World ends brilliantly with a mighty empire, the *second* is commenced by the reaction resulting from the antithesis occasioned by that infinite falsehood which rules the destinies of the *Middle Ages* and constitutes their life and spirit.

The construction of "medieval Germany" is thus considered not only the abject space to a reigning notion of the nation-state's modernity, but also a topos of baroque mysteries, hidden meanings and unexploited wealth. *JSN* implies in its opening sequence the collapse of civilization upon Charlemagne's death, and the redemption of the individual (not the collective) through finding Charlemagne's treasure in 2008. Bettina Bildhauer writes that the common assumptions about the Middle Ages include the populace being less reliant on linear time, on written discourse, and on individuality than so-called "modernity."[46] These prejudices can be seen in *JSN*'s interpretation of Charlemagne: he thinks of 1,000 years into the future rather than the present of the ninth century with regard to his treasure, he has letters engraved rather than written, and his own legacy is subsumed into that of modern Germany. Very little breath is actually wasted on the history *per se*, as the emotion-system at work in the film has too much redundant information to communicate to waste viewers' time with the gritty details of medieval interpretation. One way to look at the film less cynically, of course, is to say such adventure stories awaken the desire in younger viewers to learn about the past. But Duara's school of thought would question whether or not these viewers would be viewing the German cultural heritage as a mere means to an end. The film certainly conveys the *moods* of historical exploration: from avarice for rewards, to curiosity about where the next clue is, to awe at the encounter with the various tombs and other hallowed 1,000-year-old spaces. Yet such mood-manipulation does not convey an actual encounter with the medieval—the fantasies, lifestyles, craftsmanship and worldviews of Franconia 1,000 years ago—but rather with the persistent needs of the present to crack the next code and beat the rival archaeologists to the next site. It is this utter lack of insistence on the materiality and idiosyncratic logics of that time period that renders *JSN* mute on the subject of medieval artifacts. This adventure film illustrates the present's long-term commodification processes that seek to domesticate all that is "medieval" into an implausible-yet-immersive story world that can neither acknowledge the polysemy inherent in European history nor perceive the world beyond its relatively provincial national relations within a neo-liberal global system.

Concluding Remarks

This essay has discussed the TV film *Die Jagd nach dem Schatz der Nibelungen*'s confrontation with medieval history as being thoroughly instrumental in ways that reveal neo-liberal, crypto-nationalistic assumptions at the level of the institutions that produced it. I have demonstrated that the model of *National Treasure*, which ignores slavery, only begets more problems when imported into the German national context, especially in ignoring 20th-century events that may have impacted the protagonists' unimpeded physical access to a thousand years of European medieval history. With regard to sensitivity and accuracy, one might have easily adopted the position that Pugh and Ramey do: "When cinema meets history, the very meaning of 'history' appears to crumble under the pressures to translate the truth of the past into the media of the present."[47] Yet I am not here skewering *Pope Joan* or *Vision* for their admitted inaccuracies, but remarking that the emotional appeal of *JSN* specifically positions it within the conventions of the adventure genre, such that perfect pan-generational communication lets the heroes find the treasure as they instrumentalize medieval artifacts to do it. The foreignness and otherness of ninth-century Aachen or Ralswick do not appear here, save as citations of "historic" sites in Germany to visit as consumers. There is also no pan–European road trip that can take place: only white Germans preoccupied with their own history on their own soil, encountering and colonizing the heterogeneous world of objects from the Middle Ages. I would agree with Randall Halle's assessment that "the transnationalization of Europe [nevertheless] preserves a national cinema, even as it recontextualizes that cinema, creates a break, and radically changes its significance."[48] That is to say, *JSN* expresses thorough emotional insecurity about dealing with the rest of Europe or any aspect of German history with any gravity. When RTL wants a ratings boost, it pulls out the stops and engages in acts of southern Catholic, white German nationalism to rally the majority of its viewers around emotional propositions of suspense and conquest. European history becomes a game the Germans have already won, for they can "pull out" of 20th-century history, despite the fact that, as Yehuda Cohen argues:

> Forging a European national identity must rest on more than material gain. It requires a shared narrative. Undeniably, for Europeans as a whole, the shared traumatic memory of two world wars is a strong undercurrent within European societies, pushing toward the emergence of a single unified nationality.[49]

Such biographical facts have been more or less carved into the lives of the 21st-century European populace, and now it is time for entertainment cinema to help dream of a Europe not bounded by purely national symbols and, in this case, tourism industries. A German nation bound by tourism will continue to market a consumable simulacrum of itself, and this simulacrum will have to rely on aesthetic traditions (i.e., the adventure genre) with problematic roots and patterns. The border-crossings and mutual entanglements found as a feature of Charlemagne's Holy Roman Empire must be included in a German national project that is also to embrace its Turkish, Italian, Russian and Afro-German populations, as well as their shared contributions and investments in the future of Germany.

I wish to conclude by returning to a mood-cue analysis of one final sequence from *JSN* that illustrates my point. After Eik and Katharina have successfully overcome the drowning trap in the Hall of the World, the crafted model of the Zugspitze rises from the waterlogged floor and a shaft of light shines softly on it. At this point, the viewers are relieved, and the music dims so one can only hear the sloshing of the water. Suddenly, a bright flash of light bounces off millennia-old reflective surfaces and a series of cuts show it stopping at the wall. The mood of relief is accented with a moment of wonder. "Frau Berthold, you see," Eik says, and Katharina sharply corrects him: they are now on a first-name basis. Then they look with wonder at the wall where the beam has hit. "Antillian gold!" Katharina exclaims, polishing it off with her shirt, and the beam of light reflects right into the camera. With still no music, the beam of light pierces the model, and Eik looks at it in awe. "The mountain," he says, and Katharina says, "I know—I didn't think I'd ever believe in it." The audience receives redundant cues that she has decisively shifted from being a skeptic to a believer. The music begins to swell as she continues to speak, then trails off, and then she and Eik share their first real kiss. The scene guides the viewer from relief over their survival from a death trap, to a breakthrough in their treasure search to the most important breakthrough: the protagonists' heterosexual coupling. And where does the medieval lie in the sequence? The Antilian gold on the wall, a bit of decoration from a legendary eighth-century island that Katharina is able to identify instantly. A mythical national treasure can only be found by the light of other myths, according to what utility this film projects onto a medieval world bent to the whims of the modern.

Notes

1. "German Films vergibt 106.000 Euro Förderung für Kinostarts deutscher Filme im Ausland." *filmportale.de* (12 March 2010), accessed 2 May 2013.
2. Luley.
3. Comment by horn-bou, http://www.moviepilot.de/movies/die-jagd-nach-dem-schatz-der-nibelungen/comments#, accessed 4 May 2013.
4. The irony meant here is certainly not the literary device, in which the signified is the opposite of what the signifier would suggest, but rather the gaze or disposition of the jaded viewer, who has presumably "seen it all." Post-ironic cinema and television interrogate the notions of irony and sincerity and themselves, appealing both to the naïve and jaded viewer. Post-ironic cinema achieves its appeal by either playing straight an absurdly contrived situation, or riddling a sincere situation with clichés that undermine its possible coherence. Examples of the former include *Snakes on a Plane* (2006) and *Sharknado* (2013), with examples of the latter including *Atlas Shrugged* (2010) and *The Spoils of Babylon* (2014).
5. Kriemhild is obviously the Burgundian princess from the *Nibelungenlied*, and "Krimi" is German for "mystery novel," giving her name the (post-ironic) dual function of alluding to the character from the epic poem and nodding and winking at the genre fiction unfolding in the film.
6. Torner, 69.
7. Taves, 15.
8. Torner, 70.
9. Klein in Koebner, 7.
10. Seeßlen.

11. Jameson, 316.
12. *Ibid.*, 315.
13. Pugh and Ramey, 2.
14. *Ibid.*, 6.
15. Christensen, 3.
16. Cf. Heiduschke.
17. Berghahn and Sternberg, 2.
18. For more on the notion of "cinema of attractions," see Gunning. For the cinema of attractions in the context of recent German cinema, see Rentschler.
19. Lucas, Spielberg and Kasdan, 1.
20. *Ibid.*, 8.
21. *Ibid.*, 8–9.
22. *Ibid.*, 14.
23. Grant, xv.
24. Smith, 4.
25. *Ibid.*, 46.
26. Derrida, 92.
27. Hartley, 1.
28. Bildhauer, 20.
29. Parks and Kumar, 3.
30. So argues Cohen, 32.
31. See Herder, 48–50.
32. Dirlik, 1.
33. For more on the increasing gamification and transmedialization trends of the 00s (noughties), see Jenkins.
34. Here, in mentioning the "brave and clever few," I would argue that W.E.B. DuBois' Talented Tenth, or the top 10 percent that supposedly lead the other 90 percent, is ideology that lurks under the surface of this discourse.
35. Elsaesser, *Weimar Cinema and After*, 3–4.
36. Fisher and Prager, 4.
37. The tourism industry grows about 2 percent annually, according to the Bundesverband der Deutschen Tourismuswirtschaft, with across-the-board increases in both domestic and international tourism between 2–4 percent for the year 2012.
38. On the universality of the *Nibelungenlied*, see McConnell, 105.
39. Martin, 1.
40. Martin, 26, quoting Brackert, 362.
41. See the definition of "Hoard."
42. Cohen, 28.
43. Duara, "Transnationalism" 30 and Duara, *Rescuing History*, 34.
44. Bildhauer, 11.
45. Hegel, 360–365.
46. Bildhauer, 11.
47. Pugh and Ramey, 1.
48. Halle, 7.
49. Cohen, 129.

Bibliography

Berghahn, Daniela, and Claudia Sternberg, eds. *European Cinema in Motion: Migrant and Diasporic Film in Contemporary Europe*. London: Palgrave Macmillan, 2010.
Bildhauer, Bettina. *Filming the Middle Ages*. Berlin: Reaktion Books, 2011.
Christensen, Jerome. *America's Corporate Art: The Studio Authorship of Hollywood Motion Pictures*. Stanford: Stanford University Press, 2012.
Cohen, Yehuda. *The Germans: Absent Nationality and the Holocaust*. Toronto: Sussex Academic Press, 2010.
Collins, Jim. "Genericity in the Nineties: Eclectic Irony and the New Sincerity." *Film Theory—*

Critical Concepts in Media and Cultural Studies 2. Philip Simpson, Andrew Utterson and K.J. Shepherdson, eds. New York: Routledge, 2004: 160–180.

Derrida, Jacques. *Archive Fever: A Freudian Impression.* Eric Prenowitz, trans. Chicago: University of Chicago Press, 1998.

Dirlik, Arif. "Is There History after Eurocentrism?: Globalism, Postcolonialism, and the Disavowal of History." *Cultural Critique* 42 (Spring 1999): 1–34.

Duara, Prasenjit. "Transnationalism and the Challenge to National Histories." *Rethinking American History in a Global Age.* Thomas Bender, ed. Berkeley: University of California Press, 2002.

_____. *Rescuing History from the Nation: Questioning Narratives of Modern China.* Chicago: University of Chicago Press, 1995.

Elsaesser, Thomas. *Weimar Cinema and After: Germany's Historical Imaginary.* London: Routledge, 2000.

Fisher, Jaimey, and Brad Prager, eds. *The Collapse of the Conventional: German Film and Its Politics at the Turn of the Twenty-First Century.* Detroit: Wayne State University Press, 2010.

Grant, Barry Keith. *Film Genre Reader II.* Austin: University of Texas Press, 1995.

Gunning, Tom. "The Cinema of Attractions: Early Film, Its Spectator and the Avant-Garde." *Early Cinema: Space, Frame, Narrative.* Thomas Elsaesser and Adam Barker, eds. London: BFI, 1990.

Halle, Randall. *German Film after Germany: Toward a Transnational Aesthetic.* Champaign: University of Illinois Press, 2008.

Hartley, L.P. *The Go-Between.* London: Hamish Hamilton, 1953.

Hegel, G.W.F. *The Philosophy of History.* J. Sibree, trans. Mineola, NY: Dover, 2004.

Heiduschke, Sebastian. "21 October 2001: Record Ticket Sales Make *Der Schuh des Manitu* Most Successful Film in German History." *A New History of German Cinema.* Jennifer Kapczynski and Michael Richardson, eds. Rochester: Camden House, 2012.

Herder, Johann Gottfried. "Ideen zur Philosophie der Geschichte der Menschheit." 2.16.III. *Idealismus und Nation: Zur Rekonstruktion des politischen Selbstbewußtseins der Deutschen.* Bernard Willms, ed. Paderborn: Ferdinand Schöningh, 1986: 48–50.

"Hoard." *The Nibelungen Tradition: An Encyclopedia.* Winder McConnell, Werner Wunderlich, Frank Gentry, Ulrich Mueller, eds. London: Routledge, 2013.

Jameson, Fredric. "Globalization and Hybridization." *World Cinemas, Transnational Perspectives.* Natasa Durovicova and Kathleen Newman, eds. New York: Routledge, 2010: 315–319.

Jenkins, Henry. *Convergence Culture.* New York: New York University Press, 2006.

Klein, Thomas. "Der Abenteuerfilm." *Reclams Sachlexikon des Films.* Thomas Koebner, ed. Stuttgart: Reclam, 2002.

Lucas, George, Steven Spielberg and Larry Kasdan. "Raiders of the Lost Ark: Story Conference Transcript." 23–27 January 1978. Unpublished manuscript.

Luley, Peter. "Schatz der Nibelungen " bei RTL: Drachenblut tut Quote gut." *Der Spiegel.* 31 August 2008.

Martin, Bernhard R. *Nibelungen-Metamorphosen: Die Geschichte eines Mythos.* Munich: iudicium Verlag, 1992.

McConnell, Winder. *The Nibelungenlied.* Boston: G.K. Hall, 1984.

Parks, Lisa and Shanti Kumar, eds. *Planet TV: A Global Television Reader.* New York: New York University Press, 2003.

Pugh, Tison and Lynn T. Ramey. "Filming the 'Other' Middle Ages." *Race, Class, and Gender in "Medieval"* Cinema. *Lynn T. Ramey and Tison Pugh, eds. New York:* Palgrave Macmillan, 2007: 1–14.

Rentschler, Eric. "From New German Cinema to the Post-Wall Cinema of Consensus." *Cinema and Nation.* Mette Hjort and Scott MacKenzie, eds. New York: Routledge, 2000: 260–77.

Seeßlen, Georg. *Filmwissen: Abenteuer: Grundlagen des populären Films.* Marburg: Schüren, 2011.

Smith, Greg M. *Film Structure and the Emotion System.* New York: Cambridge University Press, 2003.

Taves, Brian. *The Romance of Adventure.* Jackson: University Press of Mississippi, 1993.

Torner, Evan. "The German Adventure Film." *World Cinema Directory Germany Vol. 1.* Michelle Langford, ed. London: Intellect Books, 2012: 69–71.

Part 2

Narrative and Genre

Episodic Arthur

Merlin, Camelot *and the Visual Modernization of the Medieval Literary Romance Tradition*

MELISSA RIDLEY ELMES

Much of the scholarship devoted to modern visual presentations of the Arthurian legend focuses on film, which is the visual art form most privileged by western societies in the 21st century. But television is a medium more compatible with the medieval romance tradition that first gave rise to our contemporary ideas of King Arthur and the Knights of the Round Table. Anyone who has read more than one medieval romance might note how closely modern television programs follow that genre in terms of the basic structure of individual narrative installments. Television serials and medieval romance texts share several additional aspects: individual episodes interlace to form an overarching narrative unity; original material is adapted for different audiences; characters are introduced, altered, combined, and omitted to suit various storylines; and multiple writers are involved in their crafting. I begin this essay with a definition of the terms "episode" and "episodic" and a brief discussion of the most common features of medieval romance, moving from there to a comparison of the medieval Arthurian tradition as represented by Sir Thomas Malory's *Morte Darthur* (1485)—widely considered to be the summation and the standard for Middle English romance—with two contemporary Arthurian sitcoms—the BBC's "Merlin" (2008–2012) and the Starz channel's "Camelot" (2011). This study shows that the episodic nature of television allows modern audiences to experience a much more authentic representation of these stories as they were circulated and constructed in the Middle Ages, rendering serial television programs such as "Merlin" and "Camelot," rather than their film counterparts, understandable as the modern visual successors of the medieval literary romance tradition.

"Episode" can refer both to a single installment within a series of interlinking narratives and to individual moments within that installment. In literary terms, an "episode" is a digression that arises incidentally and naturally from the main subject of a longer narrative.[1] By the mid-19th century, the word "episode" was also used to refer to the various stories within a larger literary tradition, such as the medieval French Reynard the Fox tales.[2] In the 20th century, the definition and use of "episode" in the literary sense was expanded to incorporate the notion of moving picture installments, such as films and, later, television programs, intended to be presented as a series; this is the most commonly used iteration of the term now. In both the literary and the visual tradition, an episode is a stand-alone story that fits within a larger set of stories or an overarching narrative. It is essentially a form of serialization. All episodes in this general sense exhibit certain common characteristics: they assume audience knowledge of the overall storyline and its main characters; they begin where the prior installment left off; they follow a set dramatic pattern akin to that of classical plays (inciting event, exposition, rising action, climax, falling action); and in place of a *denouement* they offer a cliffhanger moment that leads into the next episode. As a dramatic construction, the episode permits a story to be told intermittently over long periods of time and to continue indefinitely, with each shorter segment playing a role toward presenting the audience with the greater overall narrative in a fashion controlled by the storyteller.

In addition to referring to a serial story presented in installments, the term "episode" can refer also to specific moments within a given installment, as when the storyline switches from one character's viewpoint to another or deliberately turns from one narrative arc to a new or ongoing storyline tangentially related to the greater plot. These narrative segments permit the storyteller to offer more characters and more action than could a straightforward telling of a story, because they permit the leaving of one set of characters at a moment of tension in order to pick up a different storyline featuring other characters engaged in other events. This, in turn, generates the constant narrative variety that keeps the audience engaged. By engaging in this form of episodic storytelling, storytellers can demonstrate their skill in juggling several story arcs at once, while heightening the dramatic tension for their audience members, who are left wondering how these disparate storylines will resolve themselves and link together within the larger narrative framework.[3] This episodic structure is especially important in texts not originally intended to be read silently, but rather meant to be heard aloud or seen performed. Because the audience member of an oral or visual work cannot simply go back and re-read in order to find out what she or he has missed, or skip ahead to avoid dull passages that slow the story down, the storyteller must provide continual reminders as to who is present in the story and what she or he is trying to achieve as well as fresh material to re-engage the audience.

Finally, the term "episode" can be applied to the specific scenes that are considered central to a larger story. For example, in the Robin Hood tradition as it is passed down to modern audiences there must be an archery tournament which Robin Hood attends in disguise; Robin Hood must have a band of merry men, led by his right-hand man Little John, who live in Nottingham forest; and these men must outwit King John's forces and bring solace to the poor. For modern audiences, anything that purports to be "the whole story of Robin Hood" must include these expected scenes, which in fact are strung together from several disparate Robin Hood ballads and songs from the medieval and Early Modern periods and do not appear in any single version until the 19th century.[4] In a way, "episode" as I present it here can be viewed as a series of nesting dolls: there is the largest episode—which is the serial installment or individual tale within the overarching narrative series—in which are embedded a number of shorter episodes, or scenes, featuring the various sub-plots and character arcs; and among these shorter scenes are found the episodes specific to a given legend or tradition. This episodic formulation is how the Arthurian legend develops in both the medieval romance tradition and contemporary television programming. It is a structure that derives organically from the oral tradition in which the (largely overlapping) genres of romance and performed stories were developed. Thusly conceived, episodes permit the audience of a text—whether oral, literary, or visual in nature—to keep organized large amounts of information by providing an inherently comprehensible structure comprised of larger chunks of information divided into shorter, episodic moments and connected by a unifying figure or storyline such as that of Robin Hood or King Arthur.

Although many scholars see the roots of literary romance in classical texts, the term "romance" (*romanz*) refers in its most basic sense to stories composed in the vernacular languages of medieval Europe, as differentiated from texts penned in Latin, the official language of the Church that was learned and used by the clergy and some members of the nobility.[5] Coming from the spoken languages of the people, romance stories are generally secular and comprise subjects of interest to a population consisting mainly of individuals associated with the nobility and therefore in close proximity to or part of the textual communities that produced and disseminated these stories. Eventually, the romance stories preferred by these courtly patrons developed into a tradition of tales told in a relatively stable and predictable form: these stories took place in imaginary spaces conceived within the real countries of medieval Europe, in an unspecified time, and featured exciting adventures undertaken by chivalric knights. They often began in the oral tradition, were performed in public spaces and at court, and were only written down later and after many changes and emendations to the original story. Even when they originated as written works, as in the case of the Arthurian romances penned by Chrétien de Troyes and others, there was no guarantee that the story would not be changed in subsequent iterations and copies of the original text; in fact, such alteration was nearly certain to occur, especially when stories traveled from one culture to another, as in the case of the Perceval story from France to Germany.[6] Even when the original story was not changed it was often freely embellished, as with the French continuations to Chrétien's *Percival*.[7] Storylines and characters appeared in various new forms from one culture to the next in accordance with authorial and audience tastes in a process of adaptation that, as I will show, continues in the modern television serials devoted to the Matter of Britain.

Difficult to classify and fluid in nature, the literary romance tradition nevertheless possesses certain trademark qualities—chief among them the timeless and adventurous nature of the stories—and engages with certain specific images and motifs, especially the knight-errant, the quest, the development of the hero, the winning of the damsel, and the upholding of symbolic ideals of human behavior. It is also inherently episodic.[8] Corinne Saunders notes that romance is an open-ended genre that defies resolution—"striking in its open-endedness" if "frustrating in its capacity to defy classification or resolution."[9]—much as an episode is an open-ended installment of a story that ends on a cliffhanger rather than in a moment of resolution. This generic structure permits the storyteller and the audience alike to focus on the content rather than the format of the narrative. The underlying focus of the romance genre lies not in structural originality and the invention of new formats but rather in a self-reflexive and discursive study of timeless, trans-historical ideals of humanity that at times crosses into a form of literary nostalgia; because of this self-reflective and discursive quality the genre is, as Saunders claims, "both escapist and socially pertinent, looking backwards and forwards."[10] While remaining timeless, each romance is a product of its own time and place, making use of established characters, settings, and motifs. Despite its fluid and open-ended nature, the romance tradition thus shows itself to be a sophisticated blend of form and function achieved through, rather than in spite of, its episodic structure—much as is the television serial.

Sir Thomas Malory's *Le Morte Darthur* is the most comprehensive single-volume example of the medieval literary romance tradition.[11] Printed by William Caxton in 1485, this work benefitted enormously from Malory's unprecedented access to source materials, which permitted its being marketed and sold bearing the subtitle "The hoole book of Kyng Arthur and of his noble Knyghtes of the Rounde Table." Writing at the end of the 15th century with access to a wide variety of Arthurian chronicles and romances composed both in English and in French throughout the earlier medieval period, Malory synthesized the Arthurian legend into a sustained prose narrative that spanned the entirety of the story from Arthur's conception through his final battle.[12] Malory, himself a knight, chose to emphasize the chivalric aspects of the legend, and as W.R.J. Barron points out, this decision, coupled with Malory's comprehensive approach in including all of the Arthurian stories he was aware of, renders *Le Morte Darthur* as representative a romance as it is possible for a text to be; it "employs all the motifs of the romance genre: secret meetings between the lovers spied upon by jealous troublemakers, quest and tournament demonstrating the hero's prowess, a rival love breeding dissension, fidelity repeatedly proven in trial by combat, a bedroom siege, and rescue from the stake."[13] More specifically, the *Morte Darthur* includes all of the episodes (in the nature of "episode" as a specific scene within the larger narrative) best-associated with the Arthurian tradition, including the pulling of the sword from the stone, the love triangle between Arthur, Guinevere and Lancelot, the Grail quest, and Arthur's death. The work also features the characters—King Arthur, Guinevere, Lancelot, Gawain, Tristan, and Perceval, among others—that audiences familiar with the various cycles and iterations of the legend might expect to see. Malory's compilation and distillation of all of these figures into a single, more or less unified narrative colors subsequent audience expectations of the Arthurian legend by defining its parameters. Because it comprises all of the motifs and characters we most closely associate with Arthurian romance, Malory's *Le Morte Darthur* constitutes a single-volume example of medieval romance suitable for comparing that textual tradition with the visual tradition of Arthurian television programs; in fact, Malory's *Morte Darthur* is the most widely consulted and referenced of all of the Arthurian medieval romances and provides the backbone for most modern adaptations of the legend, including "Merlin" and "Camelot."

In the earlier tales that served as Malory's sources, multiple writers continued, expanded upon, truncated, altered, and took previous versions of the legend into different directions. Their combined efforts created a labyrinthine episodic narrative that appeared over the course of many centuries, one that privileged the individual adventures over a unified overall narrative. By contrast, Malory combines these various episodes through a process of interlacement to create internal unities, heighten dramatic tension, and provide a sense of continuity to the overall narrative of the *Morte Darthur*. In doing so, he shows himself concerned both with the structure of the story—leaving the reader with one tale at a cliff-hanging moment in order to take up another storyline—and also the subject matter, making sure to include each of the moments, or episodes, an audience familiar with the various earlier iterations of the Arthurian tradition might expect, such as Arthur's drawing of the sword from the stone and Lancelot's and Guinevere's adulterous relationship.[14] Malory also expands upon and alters the narratives of his source materials, most famously with the introduction of what seems to be an original storyline for Gareth of Orkney. By creating this additional tale to insert into the overall narrative, Malory is taking part in the time-honored activity of inventing new stories to attract and retain his audience's attention, but doing so with consideration for the new story's place within the overall narrative, as well as its entertainment value for his audience.[15] It is in Malory's hands that the medieval romance takes on the unified yet still episodic quality that contemporary audiences have come to expect in Arthurian literature and television, alike.

Malory's use of incipit and explicit passages to indicate the beginning and ending of each longer tale in the *Morte Darthur*, and the deliberate interlacement of the tales, are his most important contributions to the romance tradition that I view as culminating in the televised Arthurian series examined in this study. In particular, Malory's technique of referencing the events of earlier episodes and connecting them to the main plot through the regular presence of a cadre of individual characters reappearing in each new tale allows him both to preserve the central story, and to extend that storyline to incorporate myriad other characters and subplots into the overall narrative. Most commonly, Malory uses expressions such as "Now turne we unto"[16]; "The meanewhyle as thys was adoynge"[17]; "so leve we [...] and turne we unto"[18]; and "here this tale overlappeth a whyle unto"[19] to achieve this episodic effect in his storytelling. He employs such phrases throughout the *Morte Darthur* to indicate a break from the ongoing narrative either to start at a new place or turn to a different character in the current storyline, to follow a character to another ongoing storyline, or to begin a new story that will eventually interlace with the overall narrative arc. These cues function much as do scene changes in a television program, permitting the audience, whether listening or reading along, to follow the intricacies of his narrative. Through these techniques Malory achieves an unprecedented degree of cohesion in the medieval romance narrative that anticipates the form of the modern television serial with its central cast and storyline enhanced by guest stars and their various subplots.

Broadcast over the course of five seasons (2008–2012), the BBC's "Merlin" evinces all of the characteristics associated with medieval romance: it is, of course, filmed in the vernacular tongue; it draws on a variety of sources and takes them in new or different directions; and it employs a highly episodic narrative structure accrued incrementally over a lengthy period of time. As a modern take on the medieval Arthurian legend, "Merlin" features all of the major motifs—secret meetings, quests and tournaments, love triangles, trials by combat, bedroom sieges and near-miss executions—and major characters, including Merlin, Arthur Pendragon, Lancelot, and Guinevere, found in the *Morte Darthur.* Like Malory's *Morte Darthur*, "Merlin" features a complex interlacement of individual narrative arcs for each of the characters that feeds into an overarching series storyline; each installment follows a set pattern of events, and within each of the installments, the audience is presented with cues to help them follow different storylines toward each episode's conclusion. Because the elements and structure of this serial program so closely mirror the elements and structure we expect in a medieval romance, it is reasonable to claim both that "Merlin" is a modern visual counterpart to the medieval romance genre, and that the serial nature of television production renders it a more suitable medium for adaptations of medieval romance than the more finite nature of film, allowing as it does for the leisurely development of several storylines over a long period of time. Recognizing this permits us to understand such television adaptations as a modern experience of the romance narrative tradition that is very like what we envision the medieval audience experience as having been.

"Merlin" further mirrors the medieval romance tradition in blending traditional Arthurian tropes and expected elements of the story with material adapted from a variety of other sources, many contemporary to the show itself. The "Merlin" cast members are British and their characters speak with British accents, which is customary in programs dealing with the ultimate secular British legend: King Arthur. Although originating in the source material of the medieval Arthurian tradition—including Geoffrey of Monmouth's *Historia Regum Britanniae*, the *Prose Merlin* tradition, and Malory's *Morte Darthur*—"Merlin" is also influenced by more contemporary material, including the U.S. television program "Smallville," which deals with the early years of Superman in much the same way "Merlin" explores the early years of the eponymous character.[20] Providing a "prequel" to the traditional narrative permits the writers of "Merlin" to create new storylines that are at once fresh and original and also familiar because they provide backstory for and are therefore affiliated with the already-established legend. This is similar to the way in which medieval romance writers inserted original material and new storylines into the already-existing Arthurian timeline.[21]

Like its medieval romance counterparts, "Merlin" also presents sociopolitical issues that resonate with contemporary audiences: where Malory's work focuses on the chivalric communities and feudal system of 15th-century England, "Merlin" focuses on authorities abusing their power and questions of race and class. To do this, the writers of "Merlin" alter the medieval Arthurian legend in three essential points. First, magic is outlawed in Camelot by Uther Pendragon (played by Anthony Head), which requires Merlin (played by Colin Morgan) to mask his abilities rather than using them outright to serve the king.[22] Second, Guinevere, or Gwen (played by biracial actress Angel Coulby) is the daughter of Tom, a blacksmith, and maid to Lady Morgana (played by Katie McGrath as Uther's ward, rather than Arthur's sister[23]) where in the medieval tradition Guinevere is the royal daughter of King Lodgreaunce of Camelerde. Third, Mordred is a druid boy prophesied to bring about the downfall of Arthur with the help of Morgana, rather than on his own as Arthur's illegitimate son.[24] These three alterations to the traditional plot of the Arthurian narrative indicate an attempt to draw and capture the attention of contemporary viewers by addressing issues that are reflected in current affairs: the socio-political problems inherent in adhering too rigidly to one's principles rather than seeking common ground, characterized in Uther's often unreasonable stance against those who perform magic; issues of class (Bradley James's typecast blond, blue-eyed Arthur causes utter chaos when he defies his father in order to court—and eventually, once Uther dies, to marry—the lower-caste Gwen); and questions of race and citizenship, raised by the druids' constant persecution at the hands of Uther's army.

While maintaining the issues of honor and loyalty that are the most prevalent themes of its medieval romance counterpart, "Merlin" alters and expands upon the original source material for greater dramatic intensity. For example, Morgana's role as a magical being remains intact, but rather than Arthur's half-sister, she is presented as the daughter of the slain King Gorlois and Uther's ward, which substantially increases the amount of time she spends at Camelot. Her betrayal of Uther and Arthur—constantly foreshadowed by the prophecies of the dragon who advises Merlin (voiced by John Hurt)—therefore takes on added emotional intensity, because she suffers greater internal conflict over her choices due to her sense of obligation and loyalty to Uther, who is the only father-figure she has ever known.[25] This conflict manifests itself in the bond Morgana forges with the druids over their shared magical powers and which causes her to struggle with her sense of justice over Uther's persecution of the druids and her affection for, and sense of duty and obligation to, Uther as his ward.[26] Equally, Uther's role as her guardian renders the dramatic intensity of his constant persecution of magical individuals more acute for the viewer, who wonders if Uther will learn of Morgana's powers and what his response will be. The constant tension presented by Morgana's manifesting of and attempts to conceal magical powers further heightens the audience's concern for Merlin himself, who from the series premiere has been hiding his powers in fear of Uther's laws, but who now wishes to reveal them in order to support Morgana. Under the reign of the magic-hating Uther Pendragon, Morgana's nascent powers and her attempts to hide and control them, juxtaposed against Merlin's growing powers and attempts to reach out to a fellow magical being without being caught, form a compelling and ongoing double-jeopardy situation that would not be present if Morgana remained the distant half-sister of the medieval romance tradition.

In addition to functioning as a foil for Merlin, Morgana as Uther's ward—and therefore under his and Arthur's protection—fulfills the role of the damsel-in-distress present in the medieval romance tradition and heightened via its filtering through the 19th-century medieval revival. She also satisfies a contemporary audience's desire for strong female characters by defying Uther and, ultimately, becoming the show's central antagonist.[27] She accomplishes this through the assistance of Morgause[28] and of the sorceress Nimueh, here characterized as an antagonist to Camelot.[29] Gwen similarly is sometimes victim to her position as a servant[30] or to dangers assumed through her role as Morgana's servant, as when she is mistaken for Morgana and kidnapped for ransom,[31] and sometimes the one who saves the day, as when she rescues Arthur from a love enchantment.[32] The extra conflicts both women's socio-political differences present to the main (White Anglo Saxon Protestant) characters permit an expansion of their roles within the larger narrative, which in turn provides them with more (and more integrally important) screen time than a faithful adaptation of the medieval legend would offer. This centralization of female and minority characters as major, rather than minor, figures in the narrative is necessary for a contemporary audience from a marketing standpoint. It enables "Merlin" to attract and keep the attention of male viewers invested in the traditionally masculine ideals of chivalry and heroism explored by the romance tradition, while also appealing to female viewers seeking strong female characters and racially diverse viewers who, for once, find themselves mirrored in the traditionally very Caucasian Arthurian legend.[33]

Visual anachronism abounds, linking the show to its textual predecessors in its ahistorical approach to the material. Castle Camelot itself is a stark, grey flagstone building reminiscent of medieval castles, but its furnishings are a 19th-century poet's sumptuous re-imaginings of the world of Marie de France's Arthurian *lais*: curtained beds covered in silk hangings, tables spread with golden plates and goblets, grand wardrobes and cupboards holding a dizzying variety of clothes with a neo-Gothic flair. Morgana's and Gwen's gowns are modern takes on medieval styles made of chiffon, silk, and sateen, in bright colors including green, gold, red, and purple; Morgana is often presented with bare shoulders or in deep, cleavage-enhancing necklines, or in a diaphanous white, figure-revealing shift, while Gwen's gowns are much more noble and costly than would have been the case for a true medieval servant. Whenever Morgana goes outside, she is clad in a cape of green or red velvet that drapes delightfully and impractically down the flanks of her horse, but would be far too long to permit her to walk were she to dismount. Arthur is generally found either clad in a long tunic—cut to enhance his shoulders—over snug breeches with knee-high boots, walking around a room shirtless, or in the armor of Camelot's knights, which features a chain-mail shirt with a single shoulder plate under a billowing red cloak. The knights of Camelot, including Arthur, wear armor more than anything else throughout the series: while hunting, while relaxing, and while eating, as well as when on patrol and in tournaments and actual instances of battle, a phenomenon known as the "twenty-four hour armor trope."[34] For his part, Merlin wears a red neckerchief with a blue tunic over tan breeches, or a blue neckerchief with a red tunic over tan breeches, occasionally adding a highly anachronistic tan leather jacket. Gaius, the court physician and Merlin's mentor, wears spectacles that would not be out of place in a Victorian novel. Like Malory's *Morte Darthur* and other medieval romances "Merlin" aspires to appear both historical and relevant to contemporary visual preferences, resulting in anachronistic imagery that at once suggests a particular historical time and place and represents itself as ahistorical and timeless within that setting.

Despite veering so far from source materials in its conception as "the story before the story" and in its anachronistic adaptation of the traditional narrative and characters for contemporary sensibilities, "Merlin" takes pains to maintain the specific episodes most closely associated with the Arthurian legend. Accounted for within the overall series are such standard elements as the love triangle between Arthur, Lancelot, and Guinevere,[35] tournaments, jousting, and secret identities,[36] the loathly lady,[37] the Fisher King,[38] and the mortally wounded Arthur's voyage to Avalon and the return of Excalibur to the Lady of the Lake.[39] In addition to preserving these central episodes in the Arthurian tradition, the show's overall narrative arc—that Merlin, Arthur's servant, has magical powers he must keep hidden from Uther in order to save Camelot from ruin—maintains a consistent episodic structure for each individual installment: opening credits, inciting incident, rising action, climax, falling action, a conclusion that gestures toward the next episode, and ending credits. Within this structure, numerous individual scenes aid in developing the various characters and their storylines. Each of these scenes, in turn, ends either at the point at which the information the viewer needs in order to understand what is going on has been presented, or at the moment at which any further information divulged would reveal too much and either lessen the dramatic intensity of the episode's climax, or spoil its ending.

These individual episodes of "Merlin" often revisit earlier storylines in order to resolve them or to take them up anew in a fresh direction through an interlacement technique that closely mirrors that employed by Malory. For example, the love triangle featuring Lancelot is first introduced in Season 1, then reappears in seasons 2 and 4, while Nimueh appears frequently throughout the first season,[40] then disappears from the visual storyline but is continuously referenced in conversation between other characters in later seasons.[41] As with Malory's *Morte Darthur*, in "Merlin" this doubling-back to earlier characters and events creates a sense of unity and cohesion in an otherwise rambling narrative. Taking into account the various affinities traced above, "Merlin" can be compared to the medieval romance tradition with its adaptations, alterations, and introduction of new characters, and especially to Sir Thomas Malory's *Morte Darthur* with its overall interlacing narrative divided into individual tales, themselves divided into individual scenes among which are found the most central episodes of Arthurian legend. But "Merlin" can also be viewed as a visual continuation of the medieval literary romance tradition, and thus as an opportunity for modern audiences to reflect on the authenticity of their experience of the narrative in visual terms in comparison to what medieval audiences might have experienced in listening to or reading romances.

While "Merlin" retains the marvelous, anachronistic, and timeless qualities of medieval romance, the Starz channel's single-season "Camelot" (2011) positions itself as a more realistic take on the Arthurian legend, eschewing the marvelous magic of the romance in favor of magic associated with paganism and visually situating the story in the early medieval period in which a historical Arthur might have lived. However, despite this attempt to portray a grittier and more realistic Arthurian world, "Camelot" is as much a modern visual adaptation of the medieval romance tradition as is "Merlin."

Like the medieval romances, "Camelot" defies ready classification and requires its audience both to be familiar with the legend and willing to accept changes made to it in the spirit of fresh storytelling. Advance press releases classified the show as a "10-episode epic drama [that] redefines the classic medieval tale of King Arthur"[42] but the series's overview suggests both a quasi-historical take on the story as found in medieval chronicles and a darkly romanticized version of the story reminiscent of such modern works of Arthuriana as Marion Zimmer Bradley's neopagan *Mists of Avalon*:

> In the wake of King Uther's sudden death, chaos threatens to engulf Britain. When the sorcerer Merlin has visions of a dark future, he installs the young and impetuous Arthur, Uther's unknown son and heir, who has been raised from birth as a commoner. But Arthur's cold and ambitious half sister Morgan will fight him to the bitter end, summoning unnatural forces to claim the crown in this epic battle for control. These are dark times indeed for the new King, with Guinevere being the only shining light in Arthur's harsh world. Faced with profound moral decisions, and the challenge of uniting a kingdom broken by war and steeped in deception, Arthur will be tested beyond imagination.[43]

The advance review and series overview present different expectations of the show. The former classifies it as an "epic," suggesting a program focused on the action and heroic deeds associated with nation-building, while the latter, although also using the term "epic" to describe the material, focuses predominantly on the relationships between the main characters and the personal struggle and cost Arthur will undergo. Rather than fitting neatly into either category, the show can better be described as a hybrid—a conflation of the qualities of epic and romance. The program thusly evinces the category crisis that often arises in attempting to classify a romance.

This notion of category crisis is further borne out in the show's presentation of the Matter of Britain as belonging to the early medieval period—the "Dark Ages"—even as it attempts to maintain the glamour associated with later iterations of the legend like the *Morte Darthur,* and to deploy the material as a vehicle for socio-political criticism.[44] Although "Camelot" seeks to set itself apart as being a more realistic and less fanciful version of the story—an "epic" retelling— its overall approach to the legend preserves and updates the material in ways similar to "Merlin," and this is because of the very nature of their common source material. Both "Merlin" and "Camelot" are derived from the medieval romance tradition more or less as it has been passed down from Malory's *Morte Darthur*, and because the *Morte Darthur* is an ahistorical, anachronistic, and socio-politically critical amalgamation of materials, so are its modern visual adaptations. In its attempts to push back against lighter-hearted modernizations, "Camelot" in fact places itself firmly in the tradition of repurposing established material to new ends that lies at the heart of the medieval romance cycle.

Evidence for this claim lies in the program's attempts at medieval realism. With the show set in the "Dark Ages," Camelot becomes a citadel sacked and ruined by Uther's enemies that Arthur and his men must laboriously restore; Morgan is sent away to live in a nunnery as a young girl; and individuals are regularly covered in dirt and grime while they are journeying and working—a sort of visual antidote to the objectionable "twenty-four hour armor" trope exercised in "Merlin." Interior scenes are ablaze with torches and candles, and Uther's residence is a replication of the hall as described in any Middle English romance. Exterior scenes are filmed in panoramic view to showcase the wild, pristine lands of an imagined pre-modern England. Chibnall states of filming in the Irish countryside that: "We want the richness and vibrancy, and we want a sense of reality to it [...]. Big and bold and cinematic and epic. It's like the country has been waiting for *Camelot* to be there because the landscapes are perfect for the Dark Ages."[45]

Yet, despite these and other attempts throughout the series to recreate a "realistic Dark Ages," and again much like the medieval romances and "Merlin,"

"Camelot" abounds with anachronism, from the way the characters speak, to the castle structure and feasts, to Morgan's neo-Gothic wardrobe and Merlin's spectacularly embellished robes. Further, several key characters depart radically from their medieval origins: Guinevere is engaged to Leontes before Arthur meets her; Morgan poisons Uther Pendragon and vows to kill her stepmother Ygrain, whom she hates; Ygrain outlives Uther and goes to live with Arthur at Camelot, where she serves as a steadying influence for the fledgling army and enters into a romance with Merlin; Gawain is a disillusioned warrior who is recruited to train Arthur's army; Vivien is a servant to Morgan, rather than a Lady of the Lake in her own right; and Merlin and Morgan are at odds with one another over her use of magic and its cost both to her and to the kingdom.[46]

Some aspects of the Matter of Britain as found in Malory's *Morte Darthur* are certainly present—specifically, Arthur's being taken at birth and fostered,[47] Morgan's role as the magical half-sister constantly attempting to undermine Arthur, and the presentation of Guinevere as Arthur's love-at-first-sight mistake. The characters most closely associated with Arthur's legend are also accounted for—Arthur (played by Jamie Campbell-Bower), Guinevere (played by Tamsin Egerton), Morgan le Fay (played by Eva Green) and Merlin (played by Joseph Fiennes) in particular.[48] While the central storylines and characters expected in an adaptation of the *Morte Darthur* are present, at its heart "Camelot" seems intent on developing the story as a vehicle for socio-political critique—in so doing, assigning to the material one of the traditional roles played by medieval romances such as Thomas Chestre's "Sir Launfal" and Malory's *Morte Darthur*.[49]

Comments from the director and cast of "Camelot" make it clear that they are aware of and exploiting this historical use of the Arthurian legend as a vehicle for socio-political commentary; in an August, 2010 interview Chibnall states that "this is an adult drama, of course [...]. The amazing thing about *Camelot* is you can talk about political pursuits," while actor Joseph Fiennes adds that "the magic really lies in the political essence of the piece [...]. The characters have very strong agendas politically, and that's what I'm excited about, where the power lies rather than sort of slaying dragons and things like that."[50] That is, the series's focus lies more in the material's ability to provide a critical lens for examining human interactions in the political sphere than on faithfully preserving the traditional narrative, and in this it aligns itself with the medieval romances that also adapted the central narrative for multiple purposes including socio-political critique.

While the series attempts visually to represent a more historical setting, there are many elements in "Camelot" that reflect a desire to look not into the Dark Age roots of the Arthurian legend but rather at a neo-Dark Age version of the story rooted in neo-Gothic ideas of the medieval, particularly concerning dark magic; as Fiennes notes in two separate interviews "There will be some dark arts and we'll see people changing shape and things disappearing,"[51] and in terms of this magic, "We're [...] introducing this idea that there is an immense cost; that if you dabble in this, that you will be physically, mentally, and spiritually drained [...]. And it's really not for the faint-hearted."[52] Eva Green supports this claim, noting that "you see people changing shapes, but it's mainly ancient magic, pagan magic, magic using nature and the forces of nature: air, water, Earth, fire. So it's not 'Harry Potter' or Walt Disney. You don't wave a wand"[53] which is more in keeping with the ideas of the neopagan Wiccan religion than with historically documented medieval pagan practices.

Although these descriptions of the magic in the show fit more closely with neopagan ideas of Dark Age magic than with the historical record of pagan magic traditions,[54] "Camelot's" Morgan as played by Green is presented more like Malory's 15th-century Morgan, who was "put to scole in a nunnery, and ther she lerned so moche that she was a grete clerke of nygromancy"[55] and who uses this skill in the dark arts more continuously in efforts to thwart her brother's attempts to forge alliances than do her earlier—and closer to the pagan tradition Fiennes and Green are claiming for her—representations.[56] Morgan's first appearance in "Camelot"[57] sets her up as the series's antagonist; returned from the nunnery to which she was banished, she poisons Uther in revenge for his taking Ygrain as a wife and sending Morgan away. Over the course of the season she reveals herself to be a practitioner of the dark arts as Malory describes her, even as she is also a rival for the throne, which is not the case in the *Morte Darthur*. This character therefore both adheres more closely to her medieval counterpart in the romance tradition than does the Morgan figure in "Merlin" and also departs from that traditional presentation in order to heighten dramatic intensity; in this case, not as Uther's ward living under the same roof as Arthur and therefore growing conflicted in her loyalties as her story develops, but as the prodigal daughter returned to claim the throne she feels is rightfully hers and operating from the start out of a deep-seeded desire for revenge.

There are two specific elements of "Camelot" beyond its ambiguous genre, anachronistic nature, and function as political commentary, that link it very closely to the medieval romance tradition. While overall the medieval legend of King Arthur rarely seeks to explain ambiguous or fantastical elements, rather demanding a willing suspension of disbelief, individual tales often do seek to present answers to their audience's questions; for example, Malory coyly directs his readers to "the Frensshe book" rather than explaining anything, while Chrétien de Troyes explains how each knight (The Knight of the Lion/Yvain, the Knight of the Cart/Lancelot, and the Knight of the Grail/Perceval) received his moniker and the *Prose Merlin* explains the source of Merlin's powers. The writers of "Camelot" similarly seek to provide some context for unexplained and fantastical elements of the legend, such as who poisoned Uther Pendragon[58] and how the "Lady of the Lake" came to be,[59] in keeping with the series' desire for a more realistic take on the legend. However, in other instances the show deliberately avoids such explanation, or offers an explanation that is as fantastic as the element being explained (for example, Morgan's necromancy as a deal with the devil).[60] The series follows the medieval romance tradition in this tendency both to reveal—following the example of Chrétien de Troyes—and in Malorean fashion, to conceal or at least maintain the ambiguity of sources and explanations for various narrative elements. While this narrative technique of revealing and concealing information is not isolated to "Camelot," what is striking is how the series deliberately employs these tactics in ways that so closely follow their use in its medieval counterparts, suggesting perhaps a more nuanced familiarity with and faithfulness to the source materials than is generally acknowledged.

"Camelot" goes farther than any other modern Arthurian retelling in terms of its adherence to elements of the medieval romance tradition in one other, very specific way: the inclusion of a rape scene not found in the original Matter of Britain. Although it is not often discussed, instances of rape occur in a large number of medieval romances and, in fact, such scenes are often central to the storyline, serving as important moments of character development for the knights involved. For all intents and purposes, and contrary to the medieval tradition surrounding their marriage, "Camelot" presents the physical consummation of Arthur's and Guinevere's passion for one another as a rape scene.[61] In the series, Guinevere is engaged to marry Leontes, a fact of which Arthur is aware. After several scenes in which Arthur's and Guinevere's interactions suggest clear mutual attraction, he begs her to meet him alone on the beach. Despite her protestations of her engagement as an impregnable roadblock to any further relationship with him, Guinevere complies with this request and does, indeed, go to the beach, where Arthur pressures her into confessing her feelings for him before pulling her into a cave where they ultimately engage in sexual intercourse "just once" as Guinevere repeats over and over.

But Guinevere attempts to resist Arthur's advances. Although she is obviously deeply attracted to Arthur, she also clearly wants to uphold her vow to Leontes. Over the course of the scene she first tells Arthur to stop pressuring her to admit she cares for him. When he kisses her she tells him "no," but then worn down by his advances her "no" quickly changes to "maybe." As he continues to press his advantage we can see her conflictual response—she tells him to "stop" and frets that "they'll know," until finally, when he asks her why she is crying, she replies with evident distress and misgiving, "because I want this." Yet, at no time does she actually say "yes"—she turns rather to "just this once"—

so that Arthur's insistence on continuing, despite her equivocation and in light of his knowledge that she is betrothed to another and wishes to uphold that vow, can be viewed as pressuring her into the act and thus constitutes an act of rape. Classifying it as such is significant because of the connection of rape to the medieval romance tradition, particularly as it relates to the development of a knight's character.

In Malory's late-medieval *Morte Darthur*, rape is viewed as a crime (except, of course, for Uther's rape of Ygrain, which following the traditional source materials is orchestrated by Merlin in order that Arthur be born). However, Kathryn Gravdal specifically classifies rape as a fundamental Arthurian motif, claiming that "the very name of Arthurian romance conjures up images of valor, courtliness, and gentility; we hardly associate courtly literature with sexual violence. Yet from the earliest stages of courtly romance, the character of Arthur is linked with the narration of rape"[62]; further, Gravdal argues, with sexual violence built into most romance stories, "rape, (either attempted rape or defeat of the rapist) constitutes one of the episodic units used in the construction of a romance."[63] Indeed, in much medieval romance, rape—particularly scenes of "staged rape" or instances of rape that are actually orchestrated by the woman—has also been read as a form of chivalric instruction, as Amy Vines has shown.[64] Importantly, in the medieval tradition it is Arthur's knights who participate in instances of rape, while Arthur himself is carefully distanced from such behaviors, so that "Camelot" signals a new iteration of the legend, one in which Arthur himself engages in a rape for reasons important both to the emergent narrative and his character's development. While this scene appears to show Arthur as a hapless victim to his passions and an unwitting participant in the *jus primae noctis* or "right of first night" tradition erroneously associated with the medieval era,[65] and Guinevere as a woman using a flimsy veil of morality as a means of masking her desire to engage in sexual union with him, viewed against the rape tradition in medieval romance the moment actually sets Arthur up for his harshest lesson—that his actions have both direct and indirect consequences that can ultimately hurt those he most wants to protect. This is a lesson he must learn in order to make the transformation from arrogant and brash young man into just and self-effacing king—in order to become the King Arthur of legend. Guinevere's role is expanded from that of Arthur's wife—usually a marginal figure in the stories—into a figure whose interaction with Arthur is instrumental in his character's development. In transforming the consummation of their passion into an extramarital affair with dramatic repercussions for the Arthur figure, the writers of "Camelot" thus contribute a new episode to one of the least-known but most important motifs of medieval romance—that of the (un)intentionally ravished maiden.

Different as they are one from the other, both "Merlin" and "Camelot" are visual modernizations of the Arthurian romance tradition. Both shows are anachronistic and ahistorical, presenting the Arthurian legend in ways calculated to appeal visually to a modern audience steeped in the enchantment of medievalism, and in this they are very like their medieval counterparts that also sought to inculcate a sense of timeless nostalgia. "Camelot," like "Merlin," begins the narrative prior to Arthur's rise to kingship and ends with the dissolution of his community. In preserving the basic chronology of Arthur's life, both are like Malory's *Morte Darthur*. Like "Merlin," "Camelot" pays homage to the traditional legend by including a number of its most salient episodes—the death of Uther, the pulling of the sword from the stone, the creation of Excalibur, and Arthur's immediate love for Guinevere—while also rewriting characters and adding, changing, and combining narrative events to provide a greater sense of unity to its presentation of the story, and it is in this way that both programs resemble Malory's *Morte Darthur* most of all. That they are so different and still retain so much in common is a testament to the fluidity of the romance tradition, and this adaptable nature of the romance, in turn, is inextricable from the episodic structure that serves as the foundational building block for each branch of the Arthurian legend. It is because of the episodic nature of the romance tradition in which it first developed that the Arthurian legend can continue to be retold in so many different ways without losing the dramatic intensity, sense of authenticity, or element of novelty and invention wherein lies its ability to attract and keep an audience.

Such serial programs are also more authentic than film adaptations in their treatment of the material of medieval romance. While, like television, films are comprised of individual scenes that work together to create the overall story, no two- or even three-hour film could recreate the incremental, episodic nature of the medieval romance as can a television serial aired over the course of months and years. In a television serial, multiple storylines and narrative arcs can be brought to satisfactory conclusions over the course of multiple episodes and seasons, well signposted episode-by-episode both through an initial review of what has occurred in earlier episodes ("previously on 'Merlin...'") and in conversations between characters that reference earlier events and hint at future ones. Where films must attract and retain an audience's attention for a finite period of time, television serials—like the medieval romance cycles—require a continuous attraction and re-attraction of the audience from one episode and season to the next, achieved through constant invention and reinterpretation of the subject matter. It is this inherently episodic nature of the television serial, its narrative structure comprised of episodes, within seasons, within series, unified by a central cast of characters and interlaced storylines, that renders it the more evident visual successor of the medieval romance genre. Many choices made by the writers of these programs—from the inclusion in "Merlin" of tropes like that of the loathly lady to the addition of a rape scene in "Camelot"—attest to their awareness of and adherence to the medieval romance tradition to which they contribute a modern visual part.

Notes

1. "Episode, n.," *OED Online*, March 2013, Oxford University Press, 30 March 2013.

2. See Edward Burnett Tylor, *Researches Into the Early History of Mankind and the Development of Civilization* (London, 1865), 11.

3. The resulting ambiguities and "tangled complexity" of the resulting tales is noted by many scholars working on medieval romance; see for example chapter two in Kathryn Gravdal's *Ravishing Maidens* (Philadelphia: University of Pennsylvania Press, 1991), esp. 45–49 and Tony Davenport's chapter "Romance—From Marie to Malory"—esp. 135 on—in *Medieval Narrative: An Introduction* (Oxford: University of Oxford Press, 2004).

4. Among others, Helen Phillips remarks upon the episodic nature of the Robin Hood legend in her introduction to *Robin Hood: Medieval and Post-Medieval* (Dublin: Four Courts Press, 2005), 9–20; and A.J. Pollard has likewise noted its episodic quality in his introduction to *Imagining Robin Hood* (New York: Routledge, 2004). I present Robin Hood here as a corollary figure to that of Arthur in terms of the episodic nature of his legend's formulation in order to show that such episodic formulation extended beyond the Arthurian tradition into other medieval narrative cycles.

5. See, for example, Elizabeth Archibald, "Ancient Romance" in *A Companion to Romance: From Classical to Contemporary*, ed. Corinne Saunders (Malden, MA: Blackwell, 2007), 10–25; Ben Edwin Perry, *The Ancient Romances: A Literary-Historical Account of Their Origins* (Berkeley: University of California Press, 1967); Jack Winkler, "The Invention of Romance" in *The Search for the Novel*, ed. James Tatum (Baltimore: Johns Hopkins University Press, 1994), 23–38.

6. For discussion of Wolfram von Eschenbach's thirteenth-century adaptation of Chrétien's *Perceval*, see Timothy McFarland, "The Emergence of the German Grail Romance: Wolfram von Eschenbach, Parzival" in *The Arthur of the Germans*, eds. W.H. Jackson and S.A. Ranawake (Cardiff: University of Wales Press, 2000), 54–68.

7. John L. Grigsby provides an overview of the Perceval Continuations in *The New Arthurian Encyclopedia*, ed. Norris J. Lacy (New York: Garland, 1991), 99–100. Berndt Bastert discusses the fourteenth- century *Rappolsteiner Parzifal*, a German version of the Perceval story featuring a literal translation of the French continuations, in "Late Medieval Summations: Rappolsteiner Parzifal and Ulrich Füetrer's Buch der Abenteuer" in *The Arthur of the Germans*, eds. W.H. Jackson and S.A. Ranawake (Cardiff: University of Wales Press, 2000), 166–180.

8. I am by no means the first scholar to note the episodic nature of medieval romance; Kathryn Gravdal, for instance, writes that "medieval romance structure depends on episodic units which recur systematically but are joined in ever-changing ways" (43). The originality of my argument lies not in the claim that medieval romance—much like any long work of fiction—is episodic, but rather in my contention that this episodic nature continues in the visual representation of romance matter in television series.

9. Corinne Saunders, "Introduction," *A Companion to Romance*, 1–9: 2. An excellent study of the problem of defining medieval romance specifically can be found in K.S. Whetter's *Understanding Genre and Medieval Romance* (Burlington, VT: Ashgate, 2008).

10. Saunders, "Introduction," 3.

11. All quotations in this essay taken from Sir Thomas Malory's *Morte Darthur* are from the Norton Critical Edition edited by Stephen H.A. Shepherd (New York: W.W. Norton, 2004) which is the most readily available critical edition for a general audience. The standard edition consulted by specialists in Arthurian studies is *Malory: Works*, 3d edition in three volumes, edited by Eugene Vinaver and revised by P.J.C. Field (London: Oxford University Press, 1990).

12. For discussion of the Arthurian texts Malory either knew or knew of and made use of, see Ralph Norris, *Malory's Library: The Sources of the Morte Darthur* (Cambridge: D.S. Brewer, 2008).

13. Barron, "Arthurian Romance," 77.

14. Malory drew together the many different Arthurian storylines to create "the whole book" of Arthur and his knights; dependent upon their literary tastes and backgrounds, medieval audiences of Arthurian romance might have known several of the individual scenes or only a few of them. In the case of the former, Malory's book demonstrates his knowledge and authority through its comprehensive approach, while in the latter case he would be introducing his readers to new material, enhancing his prestige as an author. Modern audiences on the other hand expect all of them, largely as a result of Malory's efforts to bring them together into a coherent and unified narrative.

15. As opposed to such sprawling works among his sources as the *Prose Tristan* or the *Lancelot-Grail* cycles.

16. E.g. p. 29, line 12; p. 106, line 41; also found in variant structures such as "Now turne we agayne unto" (p. 236, line 42).

17. E.g. p. 47, line 36; also found in variant structures such as "So in the meanewhyle com" (p. 50 line 27).

18. E.g. p. 208, ll. 36–37; also found in variant structures such as, "Now leve we there, and speke we of" (p.170, line 15; p. 196, line 46); "now leve we [...] and lette us speke of" (p. 214, ll. 30–31) and "here leveth of the tale of [...] and here begynnyth the tale of (p. 279, ll. 43–44).

19. E.g. p. 285, line 6.

20. Mark Sweney. "Merlin: BBC Cues Up TV and Cinema Ads," *The Guardian*, August 29, 2008, http://www.theguardian.com/media/2008/aug/29/bbc.television.

21. For discussion of this practice, see Ad Putter, "Finding Time for Romance," *Medium Aevum* 63, no. 1 (Spring 1994): 1–16.

22. By contrast, in the medieval Arthurian tradition, Merlin enchants Uther Pendragon at his request to look like Igraine's husband, the Duke of Tintagel, so that Uther can slip into Castle Tintagel and sleep with Igraine, thus begetting Arthur; *c.f.* Geoffrey of Monmouth, *History of the Kings of Britain*, trans. Lewis Thorpe (New York: Penguin, 1966), 205–208; Malory, *Morte Darthur*, 3–6.

23. The figure of Morgana, or Morgan le Fay, in the medieval tradition is ambiguous and complicated; see Raymond H. Thompson, "The First and Last Love: Morgan le Fay and Arthur" in *Arthurian Women*, ed. Thelma S. Fenster (New York: Routledge, 2000), 331–344 and, more recently, Jill M. Hebert, *Morgan le Fay, Shapeshifter* (New York: Palgrave Macmillan, 2013). In Malory's version of the story, Morgan le Fay is Arthur's sister and the wife of King Uriens of Gore; see Malory, *Morte Darthur*, 942.

24. In the medieval tradition, Mordred is the illegitimate son of Arthur and his half-sister or aunt, Morgause. When Merlin tells him that a male child born in May will cause the ultimate downfall of his kingdom, Arthur orders all of the male children born in May to be put in a boat and left to die; Mordred survives and is fostered by a stranger, coming to court at fourteen and setting into action the events that lead to Arthur's death. See Malory, *Morte Darthur*, 29–30; 35.

25. In Season 3, episode 5 ("The Crystal Cave") Uther confesses to Gaius that Morgana is his daughter, bringing the story into line with the traditional understanding of Morgana as Arthur's sister.

26. Morgana discovers her powers in Season 2, episode 3 ("The Nightmare Begins"). Much of the rest of Season 2 deals with the building of her relationship with the druids and the tensions this causes for all of the characters as Uther ramps up his campaign to stamp magic out of Camelot for good.

27. 3.12–13: "The Coming of Arthur" Parts One and Two.

28. In both the medieval tradition and the series, Morgause is Morgana's and Arthur's half-sister; in the romance tradition, she is Mordred's mother, while in "Merlin" she is the daughter of Gorlois and Vivienne, a priestess of the Old Religion, and bent on vengeance for an undisclosed reason.

29. In Malory's version Nimue is Nymue, Nynyve or Nenvye, the chief Lady of the Lake, best known for her role in Pellinor's adventure during "The Weddyng of Kyng Arthur" (pp. 66; 73–77) and for imprisoning Merlin in an underwater grotto (pp. 78–79). In "Merlin" she is like Morgause a priestess of the Old Religion, and it was through her magic that Arthur was born and Ygraine died, resulting in Uther's hatred of magic and all magical beings; banished from court, she vowed vengeance on Uther for the deaths of her friends and loved ones (1.9: "Excalibur").

30. 5.2: "The Once and Future Queen."

31. 2.4: "Lancelot and Guinevere."

32. 2.10: "Sweet Dreams."

33. The medieval Arthurian legend actually incorporates a great deal of diversity, featuring as it does the unification of so many lands and peoples under Arthur's rule, but traditionally the only knight of color expressly given a storyline rather than simply being mentioned in passing is the Saracen Sir Palomides. For an example of how the Palomides character appears within the narrative, see Malory, *Morte Darthur,* pp. 337–365.

34. "In fiction, armor is often presented as a piece of everyday attire to be worn wherever you go, like a sweater. RPGs are notoriously bad about this, with characters wearing the same armour for weeks. Often characters even sleep in full battle gear" as noted in "Twenty-Four Hour Armor," *TV Tropes*, accessed May 19, 2013, http://tvtropes.org/pmwiki/pmwiki.php/Main/TwentyFour HourArmor. This continual wearing of armor regardless of time and place has been criticized in other works of medievalism, such as John Boorman's "Excalibur" (1981), both by critics and by the actors performing in the roles. Gabriel Byrne notably remarked of his scene as Uther Pendragon raping Ygraine: "I don't care how into the part you are, there's always a part of your brain that tells you, 'You are making a complete eejit of yourself. Everybody knows that that's a cushion, and besides, you're wearing a suit of armor, and who ever heard of anybody raping anybody in a suit of armor?'" (In *Hollywood Irish*, ed. Aine O'Connor [Boulder: Roberts Rinehard, 1997]; quoted at "Excalibur," *Byrneholics Online: A Little Intoxicated by Gabriel Byrne*, accessed January 20, 2014, http://www.byrneholics.com/excalibur/.)

35. 1.5: "Lancelot"; 2.4: "Lancelot and Guinevere"; 4.2: "The Darkest Hour, Part Two"; 4.9: "Lancelot du Lac."

36. 1.2: "Valiant"; 2.2: "The Once and Future Queen"; 3.11: "The Sorcerer's Shadow."

37. 2.5–2.6: "Beauty and the Beast" Parts One and Two. Although not appearing in Malory's version of the Arthurian narrative, perhaps because of his clear desire to focus on the chivalric rather than courtly aspects of the legend, the loathly lady trope is popular in the medieval English Arthurian romance tradition. Its most famous iterations are found in Geoffrey Chaucer's "Wife of Bath's Tale" from *The Canterbury Tales*, in *The Riverside Chaucer*, ed. Larry D. Benson (New York: Houghton Mifflin, 1987), 116–122; in the anonymous late medieval "Wedding of Sir Gawain and Dame Ragnelle" in *Sir Gawain: Eleven Romances and Tales*, ed. Thomas Hahn (Kalamazoo: TEAMS, 1995), 47–80, in the ballad "Marriage of Sir Gawain" (Hahn, 362–371); and in John Gower's *Confessio Amantis*, ed. Russell A. Peck and trans. Andrew Galloway (Kalamazoo: TEAMS, 2000).

38. 3.8: "The Eye of the Phoenix." The Fisher King is a central figure in the Grail legend; his wounding leads to the creation of the Waste Land which is rejuvenated by the Grail. See Malory, *Morte Darthur,* pp. 563–564, esp. note 6. In "Merlin" the Fisher King is a magician for whom Arthur must complete a quest to prove his right to the throne.

39. 5.13: "The Diamond of the Day," part two.

40. 1.3: "The Mark of Nimueh," 1.4: "The Poisoned Chalice," 1.9: "Excalibur," and 1.13: "le Mort d'Arthur."

41. 2.2: "The Once and Future Queen," 2.8: "The Sins of the Father," 3.12 and 3.13: "The Coming of Arthur" parts one and two.

42. Sheldon A. Wiebe, "First Look: Starz' Camelot!" *Entertainment News Network*, January 4, 2011, accessed May 19, 2013, http://eclipsemagazine.com/first-look-starz-camelot/.

43. Wiebe, "First Look."

44. Citing director Chris Chibnall, who pitched the show using the term ("What if you took the guys from 'Friday Night Lights' and put them in the middle of the Dark Ages, how would they respond?") several reviews assign the term "Dark Ages" to the early medieval period in which "Camelot" is set (Kathryn Shattuck, "An Arthur Worthy of the Modern Ages," *New York Times Online*, 18 February 2011, accessed January 28, 2014, http://archive.today/TThX; Michael Logan, "Starz' "Camelot" Gets Down and Dirty With a New Take on the Arthurian Legend," *TV Guide Online*, 1 April 2011, accessed January 28, 2014, http://www.tvguide.com/news/starz-camelot-arthur-1031330.aspx; Mary McNamara, "Television Review: 'Camelot,'" *Los Angeles Times Online*, 1 April 2011, accessed January 28, 2014, http://articles.latimes.com/2011/apr/01/entertainment/la-et-camelot-20110401). The term "Dark Ages" initially coined by Italian Renaissance author Francesco Petrarch as a pejorative descriptor is largely eschewed by scholars in favor of "the early medieval era." I will be referring to the period as the "Dark Ages" in this essay because that is the

term the show's creators chose to employ, and therefore informs their approach to the visual presentation of the Arthurian world.

45. Brent Hartinger, "Starz' CAMELOT Aims to Be Smart, 'Realistic' Retelling of the King Arthur Legend," *The Torch Online*, August 16, 2010, accessed May 21, 2013, http://thetorchonline.com/. This claim of a "cinematic" quality for the show betrays the underlying tensions between television and film: "cinematic" is a term usually associated with film and suggestive of a grand, visually stunning production—a spectacle to be viewed on the big screen—rather than something to be watched on a television.

46. In the medieval tradition, Guinevere is not engaged when she meets Arthur and although Merlin warns Arthur that it will spell his demise, they enter into their relationship unencumbered (Malory, *Morte Darthur*, 62.1–66.1), Uther Pendragon dies of unspecified causes (Malory, 7.11–7.17), Arthur sends for Ygrain at Merlin's bidding to verify his parentage (Malory, *Morte Darthur*, 32.16–33.19), Gawain is Arthur's nephew (Malory, *Morte Darthur*, 63.30–37), Vivien, or Viviane, is a variant of Nimue found in the *Prose Merlin* as the figure who traps Merlin (See note 39), and there is no clear relationship between Merlin and Morgan le Fay, save that Merlin is on Arthur's side and Morgan (in the later tradition) is an antagonist.

47. Although Arthur is fostered to a commoner in "Camelot," in the medieval tradition he is sent to live with Sir Ector, a peer of the realm. See Malory, *Morte Darthur*, 6.

48. Especially noteworthy in the character list is the absence of Lancelot and, indeed, most of the named knights of Arthur's Round Table, with the exception of Kay, Pellinor and Brastias (Brascias, in Malory). Rather than Lancelot, the love triangle here features Arthur, Guinevere, and the man Guinevere is betrothed to, Leontes, a knight who has pledged himself to Arthur's service and who believes in the ideals Arthur purports to uphold. Interestingly, in the literary tradition Leontes is not a medieval figure, and, indeed, can hardly be characterized as an Arthurian one; he is a character in Shakespeare's *The Winter's Tale*.

49. "Sir Launfal" is a fifteenth-century adaptation of a story that appears to have originated with Marie de France's twelfth-century *lai* "Lanval." It is notable for presenting a particularly vitriolic critique of corruption at court in comparison with earlier versions of the tale which incorporated critique much milder in tone. See "Sir Launfal" in *Middle English Romances*, ed. Stephen H.A. Shepherd (New York: W.W. Norton, 1995), 190–218; "Sir Laundevale," *Ibid.*, 352–364; "Lanval" in *The Lais of Marie de France*, trans. Robert Hanning and Joan Ferrante (Grand Rapids: Baker Books, 1978), 105–125.

50. Hartinger, "Starz' CAMELOT." Fiennes goes so far as to allude to specific political agendas being explored by naming U.S. politician Donald Rumsfeld as one of the inspirations for his portrayal of Merlin.

51. Hartinger, "Starz' CAMELOT."

52. Maureen Ryan, "Five Reasons to Check Out the Camelot Sneak Peek" in *Huffpost TV*, February 25, 2011, accessed May 22, 2013, http://www.aoltv.com/2011/02/25/reasons-watch-camelot-starz/.

53. Ryan, "Five Reasons."

54. While few people believe in the existence of magic today, there is ample evidence to suggest that in the Middle Ages it was taken seriously and viewed as belonging to a long tradition of pagan practices. See Alan Charles Kors and Edward Peters, eds., *Witchcraft in Europe, 400–1700: A Documentary History* (Philadelphia: University of Pennsylvania Press, 2001); Bengt Ankerloo and Stuart Clark, eds., *Witchcraft and Magic in Europe* (Philadelphia: University of Pennsylvania Press, 2001); Richard Kiekheffer, *Magic in the Middle Ages* (Cambridge: Cambridge University Press, 2000).

55. Malory, *Morte Darthur*, 5.44–6.2.

56. Morgan le Fay first appears in Geoffrey of Monmouth's twelfth-century *Vita Merlini* ("The Life of Merlin") as one of the nine maidens of Avalon, renowned rather than reviled for her magical abilities; she is characterized as a benevolent healer to whom Arthur is taken following his battle with Mordred. See Mark Walker, *The Life of Merlin: A New Verse Translation* (Gloucestershire: Amberley, 2011). Thomson suggests a thirteenth-century shift in the view of Morgan as a benevolent to a malevolent presence; see "The First and Last Love," 331–332.

57. 1.1: "Homecoming."

58. *Ibid.*

59. 1.4: "Lady of the Lake."

60. 1.3: "Guinevere"; 1.4: "Lady of the Lake"; 1.7: "The Long Night."

61. 1.3: "Guinevere."

62. Gravdal, *Ravishing Maidens*, 42.

63. *Ibid.*, 43.

64. In addition to Gravdal's work see Amy N. Vines, "Invisible Woman: Rape as a Chivalric Necessity in Medieval Romance" in *Sexual Culture in the Literature of Medieval Britain*, eds. Amanda Hopkins, Robert Allen Rouse, and Cory James Rushton (Cambridge: Boydell and Brewer, 2014), 161–180; also, Vines, *Women's Power in Late Medieval Romance* (Cambridge: D.S. Brewer, 2011), esp. ch. 3.

65. Known variously as "droit du seigneur" and "druit de cuissage" the lord's right to first night supposedly entitled a feudal lord to have sexual intercourse with a bride on the first night of her marriage. Most famously exploited in the 1995 film "Braveheart" directed by and starring Mel Gibson (Los Angeles: Paramount Pictures) there is no true evidence for the practice of this right in medieval Europe. See Alain Boureau, *The Lord's First Night: The Myth of the Droit de Cuissage*, trans. Lydia Cochrane (Chicago: University of Chicago Press, 1998); Richard Utz, "Mes Souvenirs Sont Peut-Etre Reconstruits: Medieval Studies, Medievalism, and the Scholarly and Popular Memories of the 'Right of the Lord's First Night,'" *Philologie im Netz* 31 (2005): 49–59.

Bibliography

Ankerloo, Bengt, and Stuart Clark. *Witchcraft and Magic in Europe*. Philadelphia: University of Philadelphia Press, 2001.

Archibald, Elizabeth. "Ancient Romance." In *A Companion to Romance: From Classical to Contemporary*, edited by Corinne Saunders, 10–25. Malden, MA: Blackwell, 2007.

Bastert, Berndt. "Late Medieval Summations: Rappolsteiner Parzifal and Ulrich Fuetrer's Buch der Abenteuer." In *The Arthur of the Germans*, edited by W.H. Jackson and S.A. Ranawake, 166–180. Cardiff: University of Wales Press, 2000.

Boureau, Alain. *The Lord's First Night: The Myth of The Droit de Cuissage*. Translated by Lydia Cochrane. Chicago: University of Chicago Press, 1998.

Braveheart. Directed by Mel Gibson. Produced by Icon Productions, The Ladd Company. Performed by Mel Gibson, Sophie Marceau and Patrick McGoohan. Paramount Pictures, 1995.

Chaucer, Geoffrey. "Wife of Bath's Tale." In *The Riverside Chaucer*, edited by Larry D. Benson, 116–122. New York: Houghton Mifflin, 1987.

Chibnall, Chris. *Camelot*. Directed by Mikael Salamon, Stefan Schwartz, Ciaran Donnelly, Jeremy Podeswa and Michelle MacLaren. Produced by Starz Channel. Performed by Joseph Fiennes, Jamie Campbell-Bower and Eva Green. 2011.

Davenport, Tony. *Medieval Narrative: An Introduction*. Oxford: Oxford University Press, 2004.

Excalibur. n.d. http://www.byrneholics.com/excalibur/ (accessed January 20, 2014).

Gower, John. *Confession Amantis*. Edited by Russell A. Peck. Vol. 1. Kalamazoo: TEAMS, 2000.

Gravdal, Kathryn. *Ravishing Maidens: Writing Rape in Medieval French Literature and Law*. Philadelphia: University of Philadelphia Press, 1991.

Grigsby, John L. "Continuations of Chretien de Troyes's Perceval." In *The New Arthurian Encyclopedia*, by Norris J. Lacy, 99–100. New York: Garland, 1991.

Hartinger, Brent. "Starz' CAMELOT Aims to Be Smart, 'Realistic' Retelling of the King Arthur Legend." August 16, 2010, http://thetorchonline.com/ (accessed May 21, 2013).

Hebert, Jill M. *Morgan le Fay, Shapeshifter*. New York: Palgrave Macmillan, 2013.

Kiekheffer, Richard. *Magic in the Middle Ages*. Cambridge: Cambridge University Press, 2000.

Kors, Alan Charles, and Edward Peters. *Witchcraft in Europe, 400–1700: A Documentary History*. Philadelphia: University of Pennsylvania Press, 2001.

"Lanval." In *The Lais of Marie de France*. Translated by Robert Hanning and Joan Ferrante, 105–125. Grand Rapids: Baker Books, 1978.

Logan, Michael. "Starz' 'Camelot' Gets Down and Dirty With a New Take on the Arthurian Legend." April 1, 2011, http://www.tvguide.com/news/starz-camelot-arthur-1031330.aspx (accessed January 28, 2014).

"Marriage of Sir Gawain." In *Sir Gawain: Eleven Romances and Tales*, edited by Thomas Hahn, 362–371. Kalamazoo: TEAMS, 1995.

Mcfarland, Timothy. "The Emergence of the German Grail Romance: Wolfram von Eschenbach,

Parzival." In *The Arthur of the Germans*, edited by W.H. Jackson and S.A. Ranawake, 54–68. Cardiff: University of Wales Press, 2000.

McNamara, Mary. "Television Review: 'Camelot.'" April 1, 2011, http://articles.latimes.com/2011/apr/01/entertainment/la-et-camelot-20110401 (accessed January 28, 2014).

Merlin. Directed by Jeremy Webb, Alice Trougton, David Moore, Justin Molotnikov and Ashley Way. Produced by BBC. Performed by John Hurt, Colin Morgan and Bradley James. 2008–2012.

Monmouth, Geoffrey of. *History of the Kings of Britain.* Translated by Lewis Thorpe. New York: Penguin, 1966.

Norris, Ralph. *Malory's Library: The Sources of the Morte Darthur.* Cambridge: D.S. Brewer, 2008.

O'Connor, Aine, ed. *Hollywood Irish.* Boulder: Roberts Rinehart, 1997.

Oxford University Press. *"Episode," n.* March 30, 2013. http://www.oed.com/view/Entry/63527?redirectedFrom=episode.

Perry, Ben Edwin. *The Ancient Romances: A Literary-Historical Account of Their Origins.* Berkeley: University of California Press, 1967.

Phillips, Helen. *Robin Hood: Medieval and Post-Medieval.* Dublin: Four Courts Press, 2005.

Pollard, A.J. *Imagining Robin Hood.* New York: Routledge, 2004.

Putter, Ad. "Finding Time For Romance." *Medium Aevum* (Spring 1994): 1–16.

Ryan, Maureen. "Five Reasons To Check Out the Camelot Sneak Peek." February 25, 2011, http://www.aoltv.com/2011/02/25/reasons-watch-camelot-starz/ (accessed May 22, 2013).

Saunders, Corinne. "Introduction." In *A Companion to Romance: From Classical to Contemporary*, by Corinne Saunders, 1–9. Malden, MA: Blackwell, 2007.

Shattuck, Kathryn. "An Arthur Worthy of the Modern Ages." February 18, 2011. http://archive.today/TThX (accessed January 28, 2014).

Shepherd, Stephen H.A. *Sir Thomas Malory's Morte Darthur.* New York: W.W. Norton, 2004.

"Sir Laundevale." In *Middle English Romances*, edited by Stephen H.A. Shepherd, 352–364. New York: W.W. Norton, 1995.

"Sir Launfal." In *Middle English Romances*, edited by Stephen H.A. Shepherd, 190–218. New York: W.W. Norton, 1995.

Sweney, Mark. "Merlin: BBC Cues Up TV and Cinema Ads." August 29, 2008, http://www.theguardian.com/media/2008/aug/29/bbc.television.

Thompson, Raymond H. "The First and Last Love: Morgan le Fay and Arthur." In *Arthurian Women*, edited by Thelma S. Fenster, 331–344. New York: Routledge, 2000.

Twenty-Four Hour Armor. n.d., http://tvtropes.org/pmwiki/pmwiki.php/Main/TwentyFourHourArmor (accessed May 19, 2013).

Tylor, Edward Burnett. *Researches into the Early History of Mankind and the Development of Civilization.* London: John Murray, 1865.

Utz, Richard. "Mes Souvenirs Sont Peut-Etre Reconstruits: Medieval Studies, Medievalism, and the Scholarly and Popular Memories of the 'Right of the Lord's First Night'." *Philologie im Netz* 31 (2005): 49–59.

Vinaver, Eugene, and P.J.C. Field. *Malory: Works*, 3d Ed. London: Oxford University Press, 1990.

Vines, Amy N. "Invisible Woman: Rape as a Chivalric Necessity in Medieval Romance." In *Seual Culture in the Literature of Medieval Britain*, edited by Amanda Hopkins, Robert Allen Rouse and Cory James Rushton, 161–180. Cambridge: Boydell and Brewer, 2014.

_____. *Women's Power in Late Medieval Romance.* Cambridge: D.S. Brewer, 2011.

Walker, Mark. *The Life of Merlin: A New Verse Translation.* Gloucestershire: Amberley, 2011.

Wallace, Randall. *Braveheart.* Directed by Mel Gibson. Produced by Icon Productions, The Ladd C. Performed by Mel Gibson, Sophie Marceau and Patrick McGoohan. Paramount Pictures, 1995.

"Wedding of Sir Gawain and Dame Ragnelle." In *Sir Gawain: Eleven Romances and Tales*, edited by Thomas Hahn, 47–80. Kalamazoo: TEAMS, 1995.

Whetter, K.S. *Understanding Genre and Medieval Romance.* Burlington, VT: Ashgate, 2008.

Wiebe, Sheldon A. "First Look: Starz' Camelot!" January 4, 2011, http://eclipsemagazine.com/first-look-starz-camelot/ (accessed May 19, 2013).

Winkler, Jack. "The Invention of Romance." In *The Search for the Novel*, by James Tatum, 23–38. Baltimore: Johns Hopkins University Press, 1994.

Are You Kidding?

King Arthur and the Knights of Justice

SANDY FEINSTEIN

In 2011, Andrew B.R. Elliott made a case for reconsidering the treatment of television shows, which had, he argued, been treated as "ephemeral," and, therefore, implicitly shallow, by contrast to assumptions about cinema, whose claim to legitimacy was based on its trade in "serious issues."[1] Recognizing the changed landscape of television today, he points to "the migration of screenwriters and stars from the big- to the small-screen; the scaling-up of landmark series and high-impact productions; technological developments," which he sees as eliding "the major differences in budget and production values" and the "validity of such divisions placed on the two formats."[2] Emphasizing "format and characterization, and their effect on the recreation of Arthurian themes," Elliott focuses on three examples of live-action Arthurian serials, namely, the post-war big screen, *The Adventures of Sir Galahad* (U.S.), and two small screen ones, the mid–20th-century *The Adventures of Sir Lancelot* (U.S.), and the early 21st-century *King Arthur's Disasters* (Great Britain).[3] Modeling his approach on that which has been used for film, his expressed intent is to address the ways these works differ from medieval narratives and conventions.[4]

Adaptations of works that focus on Arthurian characters and adventures are so numerous, though, that Elliott's focus on these three works primarily offers a comparative case study revealing different interpretations and approaches that inform television narratives of King Arthur.[5] The many differences, which Elliott acknowledges as being greater than those he considers in the article, ultimately problematize attempts to generalize about reception, authenticity, even quality. That may be why he reads the serial form in light of its Arthurian literary structure, seeing it as enacting the return of the king and renewal. Bringing back characters in new adventures, he concludes, makes serials durable. As he puts it, "inherent in the process of serialization is a process of rejuvenation and renewal" that "returning to that world" itself argues against its being "'disposable,' a one-time evocation of a medieval world in film."[6] Elliott's focus, then, is on how serialization itself enacts the once and future king, literally demonstrating its message of a return with each new episode.

Elliott's argument that the return of a new episode enacts the archetypal return of the king is well taken. Therefore, that is the starting point of my discussion, the enactment of this motif, and others, in one animated serial, *King Arthur and the Knights of Justice.*[7] More specifically, however, I will argue, Arthurian motifs are "remediated" in the serial, the term used by Jay David Bolter and Richard Grusin to refer to "the representation of one medium in another."[8] The resulting interplay of media, in turn, works actively to involve the audience. In addition, one source of the serial, Mark Twain's novel *A Connecticut Yankee in King Arthur's Court*, provides a *tour de force* model for this kind of deliberate remediation.

My close reading of both texts, the animation and Twain's novel, responds to media theorist Jason Mittell's exhortation regarding "the need for detailed specificity, not overarching generalities, in exploring media genres"[9] with the caveat that "we can never know a genre's meaning in its entirety or arrive at its ultimate definition because this is not the way genres operate" and, that, therefore, "genre definitions are always partial and contingent, emerging out of specific cultural relations, rather than abstract textual ideals."[10] Mittell's concern is with understanding genre in light of "cultural power" and "larger contexts of understanding," ultimately, to "better our understanding of how media are imbricated within their contexts of production and reception and how media work to constitute our vision of the world."[11]

The development of fandom that begins in the late 19th century informs the construction of the animation's vision of the world and the genres used to represent it. The means by which the animation—and its sources—remediate the telling of King Arthur's adventures show medievalizing geared to popular audiences and keen authorial awareness of what makes the medieval story matter across time and place.

One of Mittell's five core principals of genre analysis seem to refer to fandom when the author speaks of how "ongoing series, especially serials" feature "more active practices of fan involvement." He says:

> Audience practices of genre consumption and identification also seem to be different for television, featuring more active practices of fan involvement with ongoing series, especially serials. While we should not essentialize television's medium-

> defining practices, we need to account for the specific ways in which it operates at a particular moment.[12]

The early 1990s animated serial *King Arthur and the Knights of Justice* comes at a time before the internet facilitated fan involvement. Twain's novel, by contrast, comes at the beginning of "fandom," specifically its definition and media constructions involving sport and celebrity. The word "fandom" first appeared in late 19th-century America; in the 1990's the nature of fan involvement was in flux. Fan involvement itself radically changes in the 40 years following broadcast of the serial, and those changes impact how *King Arthur and the Knights of Justice* represents its vision of the medievalist and modern world, partly accounting for its internal and external remediations, including a spin-off video game and comic book.

Therefore, I will begin by reviewing some of the broad changes in fandom that inform the two works. In the 19th century, fandom referred to fans of baseball and theatre celebrities. Twain's novel, *A Connecticut Yankee in King Arthur's Court,* was written at a time when Sarah Bernhardt was an international theatre celebrity, baseball was becoming the national sport, and the medieval revival was in full swing in both England and the United States.[13] Yet when Hank Morgan asks someone where he is and hears, "Camelot," he remarks, "I don't seem to remember hearing of it before. Name of the asylum, likely."[14] That Hank initially assumes Camelot to be the name of an asylum—meaning both institution for the insane and refuge—is prescient when it comes to fandom as it was to develop from the 19th century to the present. Twain uses remediation to expose the "asylum" in which his narrator Hank Morgan finds himself. For Twain, and his illustrator Daniel Carter Beard, that asylum is a parodic example of injustices and barbarities perpetuated in history and then idealized by Victorian medievalism. Twain targets fans of medievalism, fans of the literary and theatrical celebrities that contribute to perpetuating cultural myths, and fans of athletes who will do anything to win. In the novel, these objects of fandom become part of the Arthurian story through internal remediation: Beard's illustrations turn the celebrities of the day into members of Arthur's court, and Hank Morgan creates two baseball teams of knights and sovereigns, which he identifies in the equivalent of lineup cards.[15] Thus, Twain and Beard remediate to satiric purpose while involving contemporary fans of baseball, drama divas, and medievalism.

According to Francesca Coppa, "fandom," the ultimate form of fan involvement, began with theatre and sports in the early 20th century,[16] though the word, in fact, appears earlier, specifically in late 19th-century recaps of baseball games. Late 20th-century examples of fandom, by contrast, involve appropriation of fictional characters and their stories, what Twain's narrator assumes he is witnessing in the Camelot "asylum." A century later, fans dress as particular characters, create new scripts for those characters, and critique each other's scripts as well as their source. This kind of participatory involvement began with the futuristic worlds of science fiction, and, not long after, embraced medievalist fantasy television serials and films. The animated television serial, *King Arthur and the Knights of Justice,* was broadcast twenty years before the popular medievalist HBO series *Game of Thrones* and eight years before Peter Jackson's *Lord of the Rings* trilogy, in short, before the internet radically influenced how fans interact with one another, and, as a consequence, impacted the expectations of fan involvement itself.[17] The animated *King Arthur and the Knights of Justice* (1992–93) aired during the transition from one kind of fandom to another, each of which took different forms based on the available technologies: Twain's satiric story and Beard's pointed illustrations take on the popularity of medievalism, baseball, and celebrities in the 19th century; the animation, by contrast, counts on viewers' interest in football players and the tropes of medievalism but airs before the internet's explosive impact on fan involvement.

Like Twain's novel, the animated series uses a sport, namely American football, to involve its viewers.[18] Remediating music and lyrics, supplemental animated shorts, and visuals invite the audience to join the football heroes in their adventures in King Arthur's Court. These are some of the calculated "active practices" specific to this early 1990's television serial for children. The active practices that rely on different media—song, montage, supplemental video tags—have been identified as one manifestation of the desire for "immediacy," defined by Bolter and Grusin as "the belief in some necessary contact point between the medium and what it represents."[19] They argue that the drive to immediacy paradoxically results in "hypermedia," which "expresses itself as multiplicity," as more and different media forms.[20] Hypermedia may be at work within a discrete, bounded text, for example in the cinematic use of montage, aural and visual, or it may extend beyond the boundaries of a specific text, with all its mediations, into new forms of the same work, as in a comic book or game. In *King Arthur and the Knights of Justice*, hypermedia expressed internally and externally, and specifically in their interplay, work toward increased viewer involvement.

The paradox of immediacy manifested by hypermedia is implicit in the prefixes of the two words. The prefixes of the terms seen by Bolter and Grusin in opposition, "hypermedia" and "immediacy," might prepare us for the "logic of hypermediacy" that "acknowledges multiple acts of representation and makes them visible."[21] While the meaning of the prefix "hyper" is obvious, with its etymology carried over intact from the Greek as "over, beyond, over much, beyond measure," the "im" of immediacy is less so because it encompasses two seemingly irreconcilable meanings: it is a negative, meaning "not," as in "not mediated"; and it is a Latin preposition, "in," suggesting something being part of the process of the "mediacy" that is mediation.[22] In other words, "immediacy" means not subject to intercession or being mediated while intercession and mediation are, nevertheless, implicit. Etymologically, immediacy is vexed, suggesting two things at once. This way of looking at immediacy, as both not being mediated and as being implicitly mediated, embraces its common association with direct contact as well as with what Bolter and Grusin call its "opposite number" or "alter ego," hypermedia, with its "multiple acts of representation."[23]

The television serial *King Arthur and the Knights of Justice* demonstrates this doubleness of immediacy through media that try to manipulate the viewer's involvement with each episode. On the one hand, the driving beat of music that may result in bouncing on a chair is still not the action of riding a horse. On the other, the visual focus on a crystal ball that dissolves into a scene of a football game directs the eye to the magical orb where a football team appears, but, nevertheless, cannot alter the fact of the viewer looking in from the outside. Each medium attempts to make direct contact between the animated action and the viewing audience, while simultaneously providing a safe distance from the content. This vexed relationship to immediacy may be seen internally through the serial's use of old and new technologies, including song, animation, sound effects, and, externally, through its print and electronic spin-offs. The hypermedia of the series used to involve its viewers work toward effecting the immediacy of participation, whether from singing along or seeing with the eyes of a magician, while still keeping King Arthur, his knights, and actions associated with his court, at a distance that mediation makes inevitable. Constant remediation and its consequences define Arthurian medievalism, enabling each version of the legend to represent a unique, contemporary "vision of the world." As a past world continually adapted, the doubleness of immediacy is particularly resonant: as new characters are brought "in" to that world through magic or accident—evoking, perhaps, the most medieval of mediators or intercessors (God)—actively engaged viewers temporarily erase mediation when made momentarily oblivious of its boundaries, which can occur when they assume the role of mediator themselves, the latter being what fandom's role playing games do.

In the animated serial *King Arthur and the Knights of Justice*, as in *A Connecticut Yankee in King Arthur's Court*, where "contemporary" characters find themselves unexpectedly and undesirably in a medieval world, the creators seem less inclined to erase the mediation than to draw attention to the multiple media that make it possible for readers or viewers to be in two worlds simultaneously. The "modern" characters who find themselves in the "other world" of Arthur's court, in both the cartoon *King Arthur and the Knights of Justice* and Twain's novel, *A Connecticut Yankee in King Arthur's Court*, dramatize the deficiencies of the past into which they enter, not the least of which concern the seemingly limited forms of media that work against immediacy and, thus, provide reassuring life-saving, or at least convenient, barriers to the represented experience.

Juxtaposing 19th-century New England and its available media against that of the medieval world and its media, Twain's novel offers a model for using hypermedia in the animated serial. His use of multiple media to create a sense of immediacy for the late 19th-century reader sheds light on the serial's use of multiple media to effect a similar purpose, audience engagement one hundred years later. Twain, setting his tale in Victorian England, begins with an authorial persona, M.T., encountering a stranger who begins to tell him his story and then stops to hand his auditor a manuscript he has written; the unfolding of the scribal narrative specifically draws attention to antique or old-style media in contrast to the latest 19th-century forms, of which the printed novel in the reader's hand is one. Indeed, it is the old forms of media (oral performance, tapestries, a non-print culture)[24] that are implicated in the perpetuation of the injustices of the time, such as lives wasted in meaningless combat or the perpetuation of a class system. Thus, while exposing the material deprivations and comparative inferiority of Arthur's age to that of Twain's own time period, the novel's remediations (as journalism, books, illustrations, advertisements, insurance chromo)[25] expose how media contribute to the construction of, or are even responsible for, a culture's behavior, which for Twain can mean the perpetration of evil.

Twain not only rewrites the tale of King Arthur's rise and fall, he makes explicit the remediation of his source, Sir Thomas Malory's *Morte Darthur,* in relation to other forms of medieval media, be it the period's prose or tapestries—and what is lacking in general. Quotations from Malory are remediated, for example, as "good Arkansas journalism"[26] and advertising columns.[27] These remediations, Twain's novelization as well as the newspapers and advertisements introduced into Camelot by his character Hank, are, in turn, remediated by Daniel Beard's illustrations.[28] The illustrations, additionally, involve readers in a game of who's who—for example, spotting the poet Alfred Lord Tennyson as Merlin and the actress Sarah Bernhardt as Clarence.[29] The illustrations, then, allow for two kinds of recognition: those of the Arthurian characters and their story; and those of contemporary celebrities and their cultural contributions. Bolter and Grusin remark on one of these kinds of recognition in their discussion of cinematic adaptations of Jane Austen novels:

> This kind of borrowing, extremely common in popular culture today, is also of course very old. An example with a long pedigree would be paintings illustrating stories from the Bible or from other literary sources, where apparently only the story content is borrowed. The contemporary entertainment industry calls such borrowing "repurposing": to take a "property" from one medium and re-use it in another. With reuse comes a necessary redefinition, but there may be no conscious interplay

> between media. The interplay happens, if at all, only for the reader or viewer who happens to know both versions and can compare them.[30]

In Twain's novel and the animation serial's "repurposing," the interplay of media highlights the cultural contexts of their respective audiences and their different access to the Arthurian story.

Having begun the novel by identifying Malory as his source and then by directly quoting from the *Morte Darthur*, Twain juxtaposes 15th-century prose story telling with not only his own distinctive style, but with the contrasting media available to him to tell his story. The interplay between media is apparently conscious and would be hard for readers to miss, whether or not they were familiar with earlier versions of Arthurian tales before coming to Twain's novel.[31] The accompanying illustrations, remediating both Twain and Malory, also evoke other media. Recognizing Tennyson in Beard's Merlin or Sarah Bernhardt as Clarence is a reminder that internally there is a "conscious interplay between media." Twain and Beard both remediate Tennyson's narrative Arthurian poem *The Idylls of the King,* which, in turn, remediated Malory's prose *Morte Darthur.* Throughout the tradition, this most resilient of legends has marked and celebrated important moments in Britain's history and culture. In Tennyson, Arthur's rise and fall mark the end of an era of Britain's glory and new challenges to theism prompted by Bradlaugh's arguments for atheism and Darwin's for evolution; more generally in the 19th century, the Arthurian literary tradition emerged with the Gothic Architectural movement and medievalist Pre-Raphaelitism and its followers.

Beard's use of Bernhardt as a model for Clarence offers yet another kind of interplay, that between a dramatic performer and an Arthurian character without any existing associations between the two. Clarence is not based on a preexisting Arthurian character; he is a page original with Twain. Similarly, Sarah Bernhardt was not cast in any Arthurian roles or tableaus. She, however, would have been recognizable to the late 19th-century reader as a dramatic performer. Using her, a living celebrity, as a model for a medieval page, is an example of the doubleness of immediacy in Twain's novel. On the one hand, Clarence may seem familiar by virtue of the actress's fame; on the other, he participates, by association, in the contemporary reader's 19th-century theatrical culture with its avid fans. As such, representing Clarence as a famous female theatrical stage actress highlights the artificial construction of the character, and, by extension, the artificiality of the Arthurian narrative that will unfold.

Arthurian tales have undergone and continue to undergo countless "repurposings" that do not depend on any one author or tale. Remediation can be hyper media-conscious: for example, Twain's remediating of Malory into multiple contemporary media (for example, advertising placards or newspaper circulars) or Beard's illustrations remediating the Merlin of Malory's prose *Morte*

Darthur embedded in Twain, the Merlin reconfigured in Twain's novel, the Merlin of Tennyson's poetic *Idylls*, and the Merlin and Tennyson of Julia Margaret Cameron's photographs. Beard, like Twain, draws attention to this interplay by emphasizing professional performers, the poet laureate and the dramatic diva of his day. By association, characters themselves may be seen as playing specific roles rather than being "authentic": for example, when a character (the magician Merlin) is remediated as a creator (poet), magician and creator become elided: in Twain, then, the poet Tennyson is not just Merlin, that old liar and teller of tedious tales as the page Clarence describes him to Hank Morgan,[32] but also a wizard, a worker of wonders. Such interplay is one way readers become involved in the first place. It is tempting to see all Arthurian productions as combining to form one enormous media-crossing serial with multiple authors over centuries. As Karolyn Kinane astutely puts it, "The interplay of media is itself a form of medievalism," reviving or appropriating "medieval narrative forms, styles," and genres.[33] The interplay of media *is* the thing that simultaneously reinforces a sense of distance from Arthur's world and a sense of being of that world—what I have referred to as the double sense of immediacy.

The means by which contemporary characters from the animated *King Arthur and the Knights of Justice* enter Arthur's medieval kingdom is both adapted from Twain and remediated in the animated serial.[34] In Twain's novel, one character, Hank Morgan, ends up in Camelot from Connecticut literally by accident: as he explains to M.T., the mediating narrator who encounters him in England some years later, he was hit on the head by a fellow named Hercules when he was head superintendent of "the great arms-factory."[35] By contrast, in the serial, multiple characters are transported while on a bus travelling in the northeastern United States to Camelot by means of "medieval" magic used for the specific intention of saving the kingdom. This magic involves a crystal ball, now a conventional device for seeing the future, though it was Edmund Spenser, in the late 16th-century *Faerie Queene*, who ascribed to Merlin the creation of a "world of glas[s]," or mirror, in which the heroine Britomart will glimpse Artegall, the Knight of Justice.[36] *King Arthur and the Knights of Justice* begins by remediating Spenser's crystal ball and the uses to which his Merlin put it.

In the animated serial, the crystal ball facilitates shifting points of view, which create immediacy: first, viewers see from Merlin's magical viewpoint that takes them into a world more like their own—of football players rather than castles and knights. Next, the glowing blue ball dissolves to a short pan of a team of football players to linger on one player who turns and faces the audience wearing a gold football helmet seen in a close-up head shot from the back, and, then, from the front, in a gold-visored crenellated helmet. This shift in point of view puts the "audience in the shoes of the main characters, just in the way they're shot."[37] The animated crystal ball that opens each episode of *King Arthur and the Knights of Justice* works to re-absorb the viewer every week with a new adventure. In each episode, the dissolve of the crystal ball erases the boundaries between the magical past and present reality. In other words, what appears is an image of contemporary America (a football game) within a remediated image (crystal ball) remediating another image (of a "medieval" Merlin); and these remediations exemplify the distance between the magical past and the present while, simultaneously, dissolving the space between the two worlds, a visual trick unique to film and video. Merlin's magic, together with the transparency of crystal, allow for, or create, literalized illusions of immediacy. The remediating camera and the image of the crystal ball give the viewer the sight of Merlin—not his point of view or his abilities as a seer, but what he sees.

The image of the crystal ball as a remediated image that draws viewers in while they are kept out (that is the "doubleness" of immediacy) does not operate in isolation. Aural media remediate the crystal ball as well, reinforcing and modifying its visual effect. When Merlin speaks, face forward to the viewer, it is as a prophet or seer, a role associated with crystal balls in general and Merlin in particular.[38] His voice intones: "And then from the field of the future a new king will come to save the world of the past." A group of football players appears at the words, "field of the future," then, at the words "new king," the player-turned-king appears, makes a 360-degree turn as he pulls out a sword seen in close-up; the viewer's attention, directed to the sword by the close-up, is held there by visual and aural effects, namely a dramatic light flashing up and off its edge which is accompanied by a magnified sound. This metaphoric "lightning and thunder" foreshadow the storm that will serve as a portal, bringing the football players—and the viewers—to Camelot. At the end of the sequence of the flashing, resounding sword, a phalanx of armored men on horseback appears, the football player-turned-armored-knight at the head. As the group rides toward the viewer, the captain-king bears his now raised sword to the words, "come to save the world of the past." Each episode ends the way it begins, albeit with a truncated version of the same opening credit sequence: an instrumental version of the theme song, the flash and clang of sword, the knights with Arthur in the lead. The combination of music, sound effects, visuals, and lyrics, engages the multiple senses of the viewer, and, when repeated week after week, reinforces audience involvement by means of regularity and familiarity, as Mittell argues is characteristic of television serials.

The lyrics of the theme song warrant closer consideration, too, because they serve as another remediated prophecy. The song, played during a visual montage of clashing swords and neighing horses, previews key components of the series: contests between good and evil; modern heroes magically transported into King Arthur's medieval court; and the storm signifying the magical powers that bring these two worlds together. The first clause, "Ride through the storm," telescopically evokes the way the team travels to Camelot and tells the viewer to do the same, pass through the storm and join the knights. The "storm" of the song is both literal and metaphorical. The clause that follows it, "see the knights fighting evil and crime," implicitly shifts the emphasis to the storm as a metaphor for "evil and crime." The lyrics introduce a new king and the Knights of Justice as "a modern day team" in a fight against evil. Each type of prophecy sees into a future that resembles that of the viewers' own time period, suggesting that Camelot's survival depends on late 20th-century heroes. Merlin's prophecies here create immediacy by revealing the audience's viewing moment as the most important time period for saving the once heroic past from itself.

In the animated serial, prophecy heightens a sense of immediacy, a very different use than in the sources. For example, in Twain's novel, the narrator Hank Morgan prophesizes what he knows to be true, partly to increase his reputation, partly to destroy that of his competitors, partly to expose fraud[39]; in Spenser, Merlin's mirror reveals a handsome knight to the heroine and prompts her adventure to find him. These prophecies highlight political and cultural concerns of the authors in their time, namely, Spenser's Elizabethan England and Twain's 19th-century America: Britomart's marriage to Artegall allegorically represents and celebrates the beginning of a glorious royal line culminating with Queen Elizabeth; Twain uses Hank's prophecies to criticize Victorian fads and frauds involving séances and self-promoting quacks. The hypermedia prophecies of the animated serial, by contrast, try literally to absorb the viewers, involving, if not actually empowering them, when they see as Merlin sees in the crystal ball or when they join the chorus of the catchy theme song. Though the song's beat and chorus of prophecy provide privileged knowledge of what is to transpire, to become a knight offers the ultimate involvement for fans. Accomplishing that goal is the challenge, one the producers attempt to achieve through the camera's shifting point of view and, later, in a spin-off comic book and game.

The audience viewing this animated action of "knights fighting evil" is also both implied subject and witness, part of the action and outside it. As earlier audiences "heard" the *aventures*, contemporary audiences literally "see" each adventure as they simultaneously hear it. They do not need to try to see it in their mind's eye because they are invited to join the heroes through a shared portal, "the storm," which will enable them to "ride" with these "knights of the Table Round." The repeated imperative "Come on ... come on!" applies to the viewer as well as to the replacement "King Arthur." The language enjoins the viewer to participate, but the medium imposes an obstacle to full immersion. Throughout the series, the point of view will shift back and forth from Merlin to that of the "modern day team of heroes," drawing in the viewer to a point.

The opening credit sequence and theme song strive to reinforce the relationship of the viewer to the action. In both song and prophecy few details emerge about the surrogate characters who are to embody the heroic ideals sung about and foreseen; even less information is forthcoming about the antagonists who appear as gray-armored nemeses on matching gray horses after the new-made Knights of Justice ride toward the viewer. The heroes are no more physically special than the viewers. They bear regional accents that are familiarly American; and their strengths are not just those involving physical prowess, but include critical thinking, creativity, mechanical dexterity, inventiveness, and empathy, among others. By contrast, the antagonists to be defeated are minions of the evil Morgana, and some are artificially created by her. For example, one of her evil creations is Blackhawk, who, though engineered to fly, is a primitive creature, as revealed by such details as his three-toed clawed feet and his stone constitution, anomalies of ominous transfigurative technology that constitute his nature. Transfiguration is an old art practiced by Merlin historically,[40] one that in the serial is remediated to a related purpose: to save or destroy a kingdom.

Merlin's voiceover makes explicit the inevitability of Arthurian remediation: if the Camelot known by Merlin in the past "is no more," it has succumbed to time. The passage of time affects the representation of Camelot, which Merlin's vatic voice suggests with the force of his declarative statement. The representation and manifestations of its demise change with the times and the different threats to each culture: in Twain, memories of the Civil War inform the destruction of knighthood by Gatling guns and other explosive devices; in the animation, the Cold War, with its escalation in weapons and robotic development, can be seen in the arms race involving air and ground weaponry. But, unlike the destruction described by Twain, annihilation of the evil knights in the cartoon lasts only as long as a single episode: King Arthur, his Round Table Knights—the gridiron King with his Knights—and their enemies are never permanently destroyed. Technology, here, sustains both the enemy and the knights trying to triumph over evil. Weapons, then, are part of the remediation that contribute to reinventing the legend and sustaining it.

One weapon, the Gatling catapult, in particular, remediates history itself. Alluding to the history of weapons development and acquisition, the Gatling catapult combines the technology of three time periods—the medieval, 19th century, and late 20th—by combining the medieval catapult with the 19th-century Gatling gun and the 20th-century technology of the underground missile launched by it. In the first episode and later ones, the miscreants deploy a catapult, a medieval weapon that appears here in the form of a gray dragon; in the final episode, the heroes use a Gatling catapult, remediating the Gatling gun used in Twain to defeat "the chivalry of England."[41] In the scene's final contest between modern knights who create their war machines with their own hands and automatons who depend on magic for their very existence, as in all previous episodes, the handy, creative replacement Round Table Knights ultimately win. Capturing weapons and kidnapping, or hiring, weapons specialists have been part of the history of war that plays out in the final episode.[42]

The relationship among the episodes is partly accomplished through the shared iterations: the same opening sequence; the football Knights at the Round Table taking an oath that asserts their new identities with first one voice proclaiming, "I am King Arthur," followed by the others stating in unison, "We are the Knights of Justice, we pledge fairness to all, to protect the weak, and vanquish the evil"; the magical, ritual arming of the Knights and their horses; and the deployed weapons themselves. The sword Excalibur on Arthur's armor, appearing initially as a static emblem, becomes an animate sword embodying his strength when he calls upon it (as first seen in episode two, "a knight's quest"). Arthur need not draw it from the stone to become king, as he has since Malory.[43] Animation literally makes objects, like the sword and animal emblems, come alive. Once brought to life, they not only know what to do but who they serve and obey.

King Arthur and the Knights of Justice marks the end of an era in animation, and its attention to weaponry made by hand seems to anticipate that end. That the evil Morgana kidnaps the master craftsman of the Knights of Justice, Tone, insinuates the potential for bad magic to conquer good magic, the latter identified with individual human invention and craft. The producers attempted to extend the life of the show by remediating it in "spin-offs" intended to create immediacy through different kinds of engagement involving the characters in Camelot. Interestingly, these spin-offs—a comic book and a video game—also bring together old and new forms, different kinds of technologies dependent on human ingenuity and manual skill or dexterity. Though Jesse Molesworth makes a compelling case for comic books as the ultimate remediating genre, the form is nonetheless technologically old-fashioned, relying on lettering, inking, words, and pictures like the medieval medium of illuminated manuscripts.[44] On the other hand, the newer medium, a video game based on the cartoon series, enables the gamer to inhabit one of the characters and control his/her story by means of his/her own level of anticipation and manual dexterity; his/her wins and losses are so many acts of participation that may be seen not merely as a means of cashing in on the success of the show but as yet another attempt to create or reinforce a sense of immediacy through active participation. Each of these works, all called *King Arthur and the Knights of Justice,* remediate one another, adapting and readapting King Arthur's legend as unique products of particular media in their time.

While war and fighting serve as motifs throughout medievalist adaptations, in this 1990's serial, no side gains a decisive victory. Fighting does not achieve the original purpose of either magician: Merlin's intended defeat of the forces of evil besieging Camelot or Morgana's intended usurpation of Arthur's kingdom; and the replacement Knights do not find all the magic keys necessary to return to their own time and place. Weapons development escalates episode to episode, as it did in the Cold War between the United States and the Soviet Union, called by Reagan the Evil Empire. A balance of power shifted at the end of 1991 when the Soviet Union broke up; the series was broadcast in early 1992 and lasted until 1993. Developed in this early period of the 1990's, the original serial reflects this historical and cultural context.

The media culture of the early 1990's was also very different from today, predating on-demand television, DVDs of television programs, and streaming. Predating these technologies, *King Arthur and the Knights of Justice* might very well have become an artifact of the past.

The process of transference, however, from a television series on a reliable, communal schedule to individualized on-demand viewing adds another intervening medium, which, in turn, impacts immediacy understood as both estranging and involving fans: on the one hand, this 1992 animation may appear dated and archaic in its technologies, animation, and representation of "advanced" weaponry; on the other, the viewer can now exert control over access. Seen in and through the context of a technological culture twenty years older, it has elicited both nostalgia and contempt.[45] Moreover, its appearance, disappearance, and reappearance occurring over the relatively short period of six years, collapses the usual time frame for a medium to seem archaic and obsolete. Yet in the midst of its questionable relevance as a remediated version of Twain, Spenser, Malory, and Geoffrey of Monmouth, DVD technology and YouTube have helped it survive.

The serial arrived on the threshold of viewers' changing expectations for both animation and access; as such, it is an important marker on the continuum of remediated medievalism. Indeed, *King Arthur and the Knights of Justice,* in its three genres—serial, comic book, video game—highlight the unique character of Arthurian remediation in particular and medievalism in general. M.J. Toswell, explaining the process of medievalizing as originating in three perspectives, writes, "first through the sceptical modern eye of the 21st-century scholar, second (though not invariably), through the romanticizing eye of 19th-century medievalist scholarship and study that is the foundation of the medievalizing impulse in the contemporary world; and third through the variable (reaching toward 'authentic') eye of the creator(s) of the text."[46] Medievalism itself, then, makes remediation inevitable. For, as Toswell says, "texts using medievalism are never quite finished, never quite authentic, never quite tangible, never quite over."[47] The animated characters in *King Arthur and the Knights of Justice* are forever trapped as animated characters in the early 1990's, having not inspired fandom of the kind that live action medievalist fantasies have since that time.

Authenticity may, then, be less the issue than the lack of fan ownership of the kind that has become typical of fandom: watching a series, reading a comic book, and playing a game based on the characters are, as Toswell suggests, not "quite tangible" enough. Becoming the characters by dressing like them and creating new adventures for them is now the ultimate expression of fandom. Animation may elicit fond memories, as *King Arthur and the Knights of Justice* has, but it rarely rises to the kind of fan involvement inspired by live action film as seen with Peter Jackson's *The Lord of the Rings;* the 1979 animated film version of the popular Tolkien trilogy did not elicit fandom either. The animated Arthurian serial, broadcast thirteen years after the animated *Lord of the Rings,* is also significant for marking the end of a time when sports and celebrities primarily defined fandom. The achievement of immediacy has since reached a new phase, one where fans try to erase mediation by appropriating the characters constructed by media: dressing like the characters, they become them; writing new adventures for them, they re-create them. This kind of fandom tries to replace mediation with the ultimate in immediacy, becoming fictional characters, something Hank Morgan assumed was taking place in the Camelot asylum.

Though adapting and remediating Arthur's story never seem to lose their attraction, becoming the characters of the medieval legend presents a challenge: there are so many King Arthurs and Round Table Knights and Ladies from which to choose; and while becoming a football-playing King Arthur may still be possible, animation, by its nature, makes the characters seem less real than live action does, highlighting the distance implicit in the genre, no matter how hypermedia attempt to break through the wall between story time and real time—and this animation, even with its comic book and control-transferring video game, epitomizes various attempts to evade what could not be evaded, at least in the early 1990's, the doubleness of immediacy itself.

Notes

1. Andrew B.R. Elliot, "The Charm of the (Re)making: Problems of Arthurian Television Serialization," *Arthuriana* 21.4 (2011): 53. For this language, Elliott cites Rod Stoneman, "Nine Notes on Cinema and Television," in *Big Picture, Small Screen: The Relations between Film and Television*, ed. Elizabeth Sklar and Donald Hoffman (Jefferson, NC: McFarland, 2002), 97.

2. Elliott, "The Charm of the (Re)making," 54.

3. *Ibid.*, 54.

4. *Ibid.*, 54, 63–64.

5. Bert Olton, *Arthurian Legends on Film and Television* (Jefferson, NC: McFarland, 2000), provides an extensive list and summary of Arthurian television series, including *King Arthur and the Knights of Justice.*

6. Elliott, "The Charm of the (Re)making," 54.

7. *King Arthur and the Knights of Justice*, created by Jean Chalopin, first aired September 13, 1992 (Chatsworth, CA: Image Entertainment, 2010), DVD.

8. Jay David Bolter and Richard Grusin, "Remediation," *Configurations* 4.3 (1996): 338, http://muse.jhu.edu/journals/configurations/v004/4.3bolter.html.

9. Jason Mittell, "A Cultural Approach to Television Genre Theory," *Cinema Journal* 40.3 (2001): 16, http://muse.jhu.edu/journals/cinema_journal/v040/40.3mittell.html.

10. Mittell, "A Cultural Approach," 16.

11. *Ibid.*, 16.

12. *Ibid.*, 17.

13. Kim Moreland, *The Medievalist Impulse in American Literature* (Charlottesville: University of Virginia, 1996), 25, discusses Mark Twain's "intense and explicitly articulated objection to the medievalist tradition in American culture" and his "almost unwilling and frequently unacknowledged attraction to this tradition."

14. Mark Twain, *A Connecticut Yankee in King Arthur's Court*, illus. Daniel Carter Beard, ed. Bernard L. Stine (Berkeley: University of California Press, 1983), 10. All quotations will be from this edition unless otherwise noted.

15. Twain, 18 and 458, for Beard's illustration and Stine's explanatory note, and 403 for the lineup card.

16. Francesca Coppa, "A Brief History of Media Fandom," in *Fan Fiction and Fan Communities in the Age of the Internet: New Essays* (Jefferson, NC: McFarland, 2006), 42, cites the *Oxford English Dictionary* (*OED*) for the initial use of the word "Fandom" applied to sports and theatre. I could not find the particular reference in the *OED*, though I did see it repeated on various fandom sites. In any case, the *OED*'s 1903 first usage is incorrect. The word "fandom" appears at least five years earlier, specifically in baseball recaps, for example in "Pocket Reds," *Boston Globe*, July 20, 1897.

17. See Coppa, "A Brief History," 53–58, for a review of the changes exerted by the internet on fandom.

18. If measured by popularity, football had by 1990 replaced baseball as America's game. For the comparative appeal and popularity of the two sports, see Jeffrey M. Jones, "Football Remains Runaway Leader as Favorite Sport," Dec. 29, 2008, Gallup Poll, http://www.gallup.com/poll/113503/football-remains-runaway-leader-favorite-sport.aspx. The graph shows the decline of baseball's popularity compared to football from 1938 to 2004.

19. Jay David Bolter and Richard Grusin, *Remediation: Understanding New Media* (Cambridge: MIT Press, 1990), 30.

20. Bolter and Grusin, *Remediation: Understanding New Media*, 33.

21. Jay David Bolter and Richard Grusin, "Remediation," 330, and *Remediation*, 33–34, refer to the "logic of hypermediacy" that "acknowledges multiple acts of representation and makes them visible."

22. *Oxford English Dictionary Online*, s.v. "im" and "hyper."

23. Jay David Bolter and Richard Grusin, "Remediation," 330.

24. Twain, 19. Hank Morgan's description of the tapestries epitomizes the perceived limitations of visual art in King Arthur's kingdom. It is not so much a matter of immediacy, as it is of verisimilitude for the literal-minded Hank: "battle pieces, they were, with horses shaped like those which children cut out of paper or create in gingerbread, with men on them in scale armor whose scales are represented by round holes—so that the man's coat looks as if it had been done with a biscuit punch."

25. For Malory's book, see Twain, 2; for the chromo, 52–53; for the newspaper, 257–262; for advertisements, 180–182.

26. Twain, 258–259.

27. *Ibid.*, 382–383.

28. Bernard L. Stine, ed., *Connecticut Yankee*, 456, n. iv, quotes Twain to show that Beard had free reign, '"to obey his *own* inspiration.... I want his genius to be totally unhampered."' Twain also wholly approved of the illustrations, while recognizing they were "more explicitly radical" than his text.

29. Stine, ed., *Connecticut Yankee*, 458, n. 21, sees the likeness of Merlin to Tennyson as "underscoring the book's satirical stance toward literature that idealized feudal values," which, while true, is just one apparent use. As for Clarence, who looks like Sarah Bernhardt, no reason for using her as the model is given (n. 18).

30. Bolter and Grusin, "Remediation," 338–39.

31. Twain, 126–135. Hank specifically complains about his Lady's archaic style and offers Americanisms to replace what he says is repetitive and bare bones prose.

32. *Ibid.*, 25.

33. Karolyn Kinane, June 16, 2014, in a marginal comment to the author.

34. More explicit allusions to Twain's *Connecticut Yankee* appear in the second episode, "a knight's quest," that begins with one of the African American characters, Breeze, saying, "It looks like Connecticut. It even smells like Connecticut. Man I wish it was Connecticut." To which another character says, "I got some news for you. It ain't Connecticut," at which comment a third character says, "Maybe it's all a bad dream" (cf. Twain, 10, 36–37, 48). The principal departure in the two works is the difference in the characters' goals: Hank Morgan seeks to destroy knight errantry and monarchy and the football team hopes to save the medieval kingdom and its king.

35. Twain, 4.

36. Edmund Spenser, *The Faerie Queene*, in *Edmund Spenser's Poetry*, 2d ed., ed. Hugh Maclean (New York: W.W. Norton, 1982), III.ii.17–18.9. All quotations are from this edition.

37. I am quoting Dave Green talking about Steven Spielberg's influential use of camera techniques in his 1980s films, in Mekado Murphy, "Reliving Childhood Excursions," *The New York Times*, June 29, 2014. Compare to Spenser's use of the crystal ball to create an illusion of immediacy for his character, Britomart, who will take it upon herself to seek what she sees in the glass, Artegall, the handsome Knight of Justice.

38. A.O.H. Jarman, "The Merlin Legend and the Welsh Tradition of Prophecy," in *The Arthur of the Welsh: The Arthurian Legend in Medieval Welsh Literature,* ed. Rachel Bromwich, A. O. H. Jarman, and Brynley F. Roberts (Cardiff: University of Wales Press, 1991), 131–132, discusses Merlin as prophetic in both of Geoffrey of Monmouth's works, the *Historia* and the *Vita Merlini,* the latter of which, he reminds us, opens with a statement claiming that Merlin was renowned as a prophet.

39. Twain, 232–236.

40. See, for example, Geoffrey of Monmouth, *The History of the Kings of Britain,* trans. and ed. Richard M. Loomis in *The Romance of Arthur*, ed. James J. Wilhelm and Laila Zamuelis Gross (New York: Garland, 1984), 61–63, where *Merlin* makes possible the birth of Arthur when he transforms Uther into Igrayne's husband Gorlois, Ulfin into one of his attendants, and himself into another.

41. Twain, 421, and, for the quotation, 432.

42. "Tone's triumph" is the last episode in the series, which ends without a final resolution of the problems the football players were brought to Camelot to address: Arthur hasn't been freed, the football players haven't been returned to their own century, and the threat of the enemy has only been temporarily contained.

43. Sir Thomas Malory, *Merlin,* in *Malory*, ed. Eugène Vinaver (London: Oxford University Press, 1971), 7–9. I am speaking only, of course, of English language versions of this now conventional action. Malory adapted French sources for the tales, such as *La Suite de Merlin*, among others.

44. Jesse Molesworth, "Comics as Remediation: Gilbert Hernandez's *Human Diastrophism*," *ImageTexT: Interdisciplinary Comics Studies* 7.1 (2013), (accessed Dec. 30, 2013), http://www.english.ufl.edu/imagetext/archives/v7_1/molesworth/, argues that comic books surpass other media precisely because they incorporate so many media.

45. See, for example, Alan Milligan, The Gaming Intelligence Agency (accessed Dec. 27, 2012), http://thegia.psy- q.ch/sites/www.thegia.com/snes/ka/ka.html, who excoriates the series and the game; and Justin Felix, rev. of *King Arthur and the Knights of Justice: The Complete Animated Series,* posted June 17, 2010 (accessed Dec. 27, 2012), *http://www.dvdtalk.com/reviews/41301/king-arthur-the-knights-of-justice-complete-series/,* who imagines the series would satisfy a certain age group while acknowledging that some adults still feel warmly about the series.

46. Toswell, "The Tropes of Medievalism," *Studies in Medievalism* XVII (2009): 74.

47. Toswell, "The Tropes of Medievalism," 74.

Bibliography

Bolter, Jay David, and Richard Grusin. *Remediation: Understanding New Meida*. Cambridge, MA: MIT Press, 1990.

Bolter, Jay David, and Richard Grusin. "Remediation," *Configurations* 4.3 (1996): 311–358. http://muse.jhu.edu/journals/configurations/v004/4.3bolter.html (accessed June 2, 2012).
Bradlaugh, Charles. *A Plea for Atheism*. London, 1864.
Cameron, Julie Margaret. *Tennyson Reading*. 1865. Photograph. New York, Metropolitan Museum of New York. http://metmuseum.org/collection/the-collection-online/search/268705 (accessed Sept. 24, 2012).
Coppa, Francesca. "A Brief History of Media Fandom." In *Fan Fiction and Fan Communities in the Age of the Internet: New Essays*. Jefferson, NC: McFarland, 2006.
Elliott, Andrew R.R. "The Charm of the (Re)making: Problems of Arthurian Television Serialization," *Arthuriana* 21.4 (2011): 53–67.
Felix, Justin. Rev. of *King Arthur and the Knights of Justice: The Complete Animated Series*. 17 June 2010. Accessed Sept. 27, 2012. http://www.dvdtalk.com/reviews/41301/king-arthur-the-knights-of-justice-complete-series/.
Game of Thrones. Created by David Benioff and D.B. Weiss, first aired April 17, 2011. New York: HBO.
Geoffrey of Monmouth. *The History of the Kinds of Britain*. Trans. and ed. Richard M. Loomis In *The Romance of Arthur*. Ed. James J. Wilhelm and Laila Zamuelis Gross. New York: Garland, 1984.
Jarman, A.O.H. "The Merlin Legend and the Welsh Tradition of Prophecy." In *The Arthur of the Welsh: The Arthurian Legend in Medieval Welsh Literature*. Ed. Rachel Bromwich, A.O.H. Jarman, and Brynley F. Roberts. Cardiff: University of Wales Press, 1991. 117–145.
Jones, Jeffrey M. "Football Remains Runaway Leader as Favorite Sport." December 29, 2008. Gallup Poll. Accessed 25 July 2014. http://www.gallup.com/poll/113503/football-remains-runaway-leader-favorite-sports.aspx.
King Arthur and the Knights of Justice. Created by Jean Chalopin, first aired September 13, 1992. Chatsworth, CA: Image Entertainment, 2010. DVD.
King Arthur & the Knights of Justice. Enix. Super Nintendo video game. 1993.
The Lord of the Rings. Directed by Ralph Bakshi. United Artists, 1978.
The Lord of the Rings. Directed by Peter Jackson. New York: New Line Cinema, 2001, 2002, 2003.
Malory, Sir Thomas. *Merlin*. In *Malory*. Ed. Eugène Vinaver. London: Oxford University Press, 1971.
Milligan, Allan. The Gaming Intelligence Agency. Accessed Dec. 27, 2012. http://rchive.thegia.com/snes/ka/ka.html.
Mittell, Jason. "A Cultural Approach to Television Genre Theory," *Cinema Journal* 40.3 (2001): 3–24. Accessed June 2, 2012. http://muse.jhu.edu/journals/cinema_journal/v040/40.3mittell.html.
Molesworth, Jesse, "Comics as Remediation: Gilbert Hernandez's *Human Diastrophism," Image TexT: Interdisciplinary Comics Studies* 7.1 (2013). Accessed Dev. 30, 2013. http://www.english.ufl.edu/imagestext/archives/v7_1/molesworth/.
Moreland, Kim. *The Medievalist Impulse in American Literature*. Charlottesville: University of Virginia, 1996.
Murphy, Mekado. "Reliving Childhood Excursions." *The New York Times* 29 June 2014. Arts and Leisure: 15.
Olton, Bert. *Arthurian Legends on Film and Television*. Jefferson, NC: McFarland, 2000.
Oxford English Dictionary. s.v. "fandom," "hyper," "im." Accessed July 25, 2014. http://www.oed.com/.
"Pocket Reds." *Boston Globe*. July 20, 1897. p. 7. 19th Century U.S. Newspapers. Accessed July 6, 2014. http://infotrac.galegroup.com/itw/infomark/0/1/1/purl=rc6_NCNP?sw_aep=psucc.
Spenser, Edmund. *The Faerie Queene*. In *Edmund Spenser's Poetry*. 2nd ed. Ed. Hugh Maclean. New York: W.W. Norton, 1982.
Stoneman, Rod. "Nine Notes on Cinema and Television." In *Big Picture, Small Screen: The Relations between Film and Television*. Ed. Elizabeth Skylar and Donald Hoffman. Jefferson, NC: McFarland, 2002.
Tennyson, Alfred Lord. *A Variorum Edition of Idylls of the King*. New York: Columbia University Press, 1973.
Toswell, M.J. "The Tropes of Medievalism," Studies in Medievalism XVII (2009): 68–76.
Twain, Mark. *A Connecticut Yankee in King Arthur's Court*. Illus. Daniel Carter Beard. Ed. Bernard L. Stine. Berkeley: University of California Press, 1983.

Part 3

Gender and Sexuality

Television's Male Gaze

The Male Perspective in TNT's Mists of Avalon

MICHAEL W. GEORGE

Marion Zimmer Bradley's *Mists of Avalon* is perhaps the freshest rendering of Arthuriana in existence.[1] The 2001 TNT miniseries promised to bring to television medievalism that same originality.[2] According to Time Warner, the miniseries had over 30 million unduplicated viewers in its first run, with a 5.9 rating for the first episode and 4.0 for the second (the same press release has *Law and Order* at a 1.8 rating and NASCAR coverage at 5.2—the highest sports rating of the summer).[3] With such high ratings, the miniseries could have set a significant precedent challenging the traditional, androcentric television representation of the Middle Ages. Unfortunately, it did not. Nominated for a Primetime Emmy for Outstanding Miniseries, it has received critical acclaim. Hollywood.com's review claims that "it works" and that it offers "a female perspective on the Arthurian legend."[4] Tim Goodman of the *San Francisco Chronicle* admires it as "impressive," claiming that its "feminist slant" makes the Arthurian tradition better and calling it a "femme-positive adaptation.[5] Even *Entertainment Weekly* praised it.[6] Positive reviews aside, Uli Edel's adaptation masquerades as a miniseries with a female perspective, participating in and perpetuating a tradition of androcentric television medievalism.

Strong women on television participate in a trend that escalated in the 1990s, characters whom Mary Magoulick has called "power women."[7] These female characters assume the masculine hero role of the warrior fighting for good. Although scholars and critics have often praised these shows for promoting strong women as television protagonists, others have noted some disturbing issues with these characters. As Magoulick observes, the women in these shows are threatened on all sides, even in their domestic spheres. They tend to rely in some way on a powerful male figure, and they radiate intense sexuality. Magoulick argues that "in spite of their female characters' strengths, these shows contain troubling and sexist messages in images and plot lines that neither advance nor celebrate feminism."[8] Instead of fresh views of strong women, Magoulick claims that "they [these programs] gave women lead roles in the tradition of male action heroes,"[9] action heroes who "wear ridiculously impractical outfits for warriors."[10] Shows such as *Xena: Warrior Princess*, *Buffy the Vampire Slayer*, and *La Femme Nikita* popularized this female hero type during the 1990s. The warrior-woman exudes patriarchal sexuality; she is independent, yet still relies on men; and she is often excessively violent.[11]

This woman warrior is "now a staple on television and in movies,"[12] especially in television medievalism. A number of medievalist television programs have appeared, and all of them, though often having strong female characters, are overwhelmingly androcentric. The BBC *Merlin* (2008–2012) follows the title character's younger years, remaining tightly focused on Merlin and his struggles. The program does present some strong, independent women, particularly Guinevere and Morgana, but these roles are either subservient to the overarching concern of Merlin's tribulations, in the former case, or are represented as evil antagonists, in the latter. The same can be said for the short-lived Starz *Camelot* (2011), which follows the early reign of Arthur. The episodes' principal concern is Arthur, though the show does have three primary female characters: Igraine, Guinevere, and Morgan. Of the three, Igraine is the most positive. Although certainly a product of patriarchy, Igraine breaks from the mold of woman warrior, showing strength and independence without reverting to gratuitous violence. Episode 8, "Igraine," offers a rare exception, where Igraine has been captured and held in Castle Pendragon. To escape rape by her guard, she seizes his dagger and stabs him multiple times. Even then, however, she does not finish by killing him. Rather, she escapes the castle.[13] Guinevere holds the middle ground. Although she is somewhat independent, her character seems necessary because of her role in the Arthurian tradition, and she serves as Arthur's love interest. The main alteration in her character is her marriage to Leontes, Arthur's primary hero, and her sexual tryst with Arthur on her wedding day, foreshadowing, were the series to have continued, her infidelity with Lancelot.[14] In this sense, she is the conventional Guinevere. Morgan is the typical evil antagonist we expect from her character. She poisons her father,[15] uses plenty of dark magic, and participates—often actively—in violent actions, even going so far as to kill Igraine, stabbing her stepmother and ensuring she suffers a slow death.[16] As we might expect from these characters, all are sexualized, even more than in the 1990s shows that established the woman warrior character. Because it is a pay-cable channel, Starz was able to provide plenty of nudity and overt sex in the ten episodes that aired. All three of the show's primary female characters have sex multiple times on screen. Igraine and Guinevere dress modestly for most of the episodes, but both reveal plenty of skin, and Morgan's dress is often revealing. Though only Morgan acts violently, all provide stimulating sights for the primarily male audience.

The syndicated *Legend of the Seeker* (2008–2010), based on Terry Goodkind's novels, although having strong female characters in Kahlan and Cara, focuses entirely on Richard Cypher. When we do see strength in the women, it is precisely the kind of strength found in the woman-warrior programs from the 1990s. Cara is a Mord-Sith, a class of women warriors. Her devotion to Richard is unyielding, and as with warriors such as Xena, her attire marks her as overly sexualized. Having blonde hair slightly longer than shoulder-length, she wears tight red leather clothing, cut low enough to reveal cleavage and thereby appealing to the 14–35-year-old heterosexual male viewers to which these shows tend to be marketed.[17] Kahlan tends to dress more conservatively, in long gowns; yet even with this dress, her low-cut, laced bodice reveals as it conceals. Both women are attracted to Richard. Cara is infatuated throughout most of the show; Kahlan is in love. What is more, audiences can count on the action culminating in elaborately-choreographed fight scenes, most of which include both Cara and Kahlan wielding various weapons and, in very Xena-like action, vanquishing their foes.

Finally, The History Channel's *Vikings* (2013-) presents women in the same manner, concentrating attention on Ragnar Lodbrok and his quest for wealth and power. The strong woman in this series, Lagertha, is Ragnar's wife, and the narrative is based on her legend told in Saxo Grammaticus's *Gesta danorum*. A shieldmaiden in Saxo, she is more of a traditional wife in the television series.[18] She does accompany Ragnar on raids and fights beside the Viking men, but just as often she is caring for her home while Ragnar is away. Lagertha is blonde and fair- skinned. Although she rarely wears revealing clothing, the clothing she wears, especially while on raids, is tailored to reveal her body shape. Moreover, Lagertha is often sexually threatened: by a gang of rapists, Ragnar's brother Rollo,[19] and Knut,[20] all in the first four episodes.[21] These instances of sexual threat, combined with her appearance, imbue Lagertha with androcentric sexuality.

All of these examples of television medievalism do contain strong female characters, but the point of view for each program conforms to the male gaze, following a solitary male protagonist, focusing on swordplay and battles, and telling the story from the male perspective. The women in these programs, if not taking up weapons and participating in gratuitous violence (Morgan, Kahlan, Cara, Lagertha), are sexualized and made objects of male fantasy (Morgan, Kahlan, Cara, Guinevere, Igraine, Lagertha). Such catering to the male viewer via violence and sexuality conforms narrowly to Laura Mulvey's idea of the male gaze, a gaze that is not at work in Bradley's novel but is in the television miniseries.

Leslie Workman defines *medievalism* as "the continuing process of creating the Middle Ages."[22] Tom Shippey modifies this slightly to "any post-medieval attempt to re-imagine the Middle Ages, or some aspect of the Middle Ages, for the modern world, in any of many different media; in academic usage, the study of the development and significance of such attempts."[23] Bradley's novel is certainly medievalism, presenting a fresh view of Arthuriana. Victoria Sharpe says that the novel "does not raise up past patriarchy as ideal, but rather seeks to empower the female characters," continuing to assert that "Bradley was one of the first to give any consideration to the women instead of following the usual male dominated story line."[24] Karen E. C. Fuog claims that Bradley "deliberately demystifies female sexuality and thereby deconstructs the notion of female Otherness; she gives her female characters power, not only as individuals, but also through the power structures in which they work."[25] However, while the televised *Mists of Avalon* does give women plenty of screen time and lines, the point of view is neither female nor feminist. Under the guise of retelling the story from the women's perspective, the miniseries merely reinforces the male gaze in television medievalism. By weakening Bradley's perspective, Edel participates in the androcentric representation prevalent in television medievalism. This essay analyzes key features of Bradley's novel that Edel omitted, changes in characterization, and the camera's perspective to illustrate how the miniseries androcentrizes Bradley's narrative.

Mulvey's pioneering work on the male gaze enables a gendered analysis of the miniseries. Mulvey published three influential essays on the topic. In her first and probably most often cited, she uses psychoanalysis to develop her concept of the gaze. For my purposes, her third section is the most important. Here, Mulvey establishes that "the woman displayed ... function[s] on two levels: as erotic object for the characters within the screen story, and as erotic object for the spectator within the auditorium."[26] She breaks down the process of cinema gaze into three perspectives: "that of the camera as it records the pro-filmic event, that of the audience as it watches the final product, and that of the characters at each other within the screen illusion."[27] She updates her position in her 1981 article, saying that she was investigating that "in-built patterns of pleasure and identification impose masculinity as 'point of view.'"[28] In this later article, she claims that "the 'grammar' of the story places the reader, listener or spectator *with* the hero."[29] Throughout her work, Mulvey argues that the Hollywood film industry creates an erotic gaze, with the masculine looker as aggressive and powerful, and the feminine recipient of the gaze as passive, to be acted upon.[30] Lois Tyson in her introductory text on literary theory claims that "in most Hollywood films, even today, the camera eye (the point of view from which the film is shot) is male: the female characters, not male, are the objects gazed on by the camera and often eroticized as if a male eye were viewing them, as if the point of view of the 'universal' moviegoer were male."[31] Essential for this article is not the eroticism on which Mulvey focuses, although that is also apparent. Rather, it is the camera's masculine point of view. Even when multiple genders are not present in the shot—even if they are not present in the film at all—we can analyze the camera's eye from a gendered perspective. In the case of a filmed adaptation, alterations in and omissions from the source text(s) form part of the gaze, for those dictate the focus of the camera. For the TNT *Mists of Avalon* this concept is imperative. I argue that the miniseries has stripped Bradley's gynocentric point of view from the film, re-androcentrizing her Arthurian narrative.

Bradley's novel adopts a female perspective, firmly placing the reader with the major female characters in the various Arthurian stories. Setting the story in post–Roman Britain at the beginning of the Anglo-Saxon invasions, she represents the roles of Celtic women as often at odds with Roman patriarchal tradition. The main point of view of the novel is Morgaine's, taken from various aspects of the medieval Morgan le Fay, a character who can trace her lineage to Geoffrey of Monmouth.[32] Bradley's strategy was to take the most prominent women, especially antagonists, from the Arthurian tradition and rewrite the story from their perspective.[33] In the novel, Morgaine narrates most of the story, and it is her story, for at the end of the Prologue, she says, "But this is my truth; I who am Morgaine tell you these things, Morgaine who was in later days called Morgan le Fay."[34] Morgaine provides a reason for her tale: "for one day the priests too will tell it [the story], as it was known to them. Perhaps between the two, some glimmering of the truth may be seen."[35] This scenario is, as we know, precisely what happened with the Arthurian tradition.[36] From Geoffrey of Monmouth to Malory, the writers of Arthurian stories were devout Christians if not clerics themselves.[37] Nearly all of Arthuriana is from a male perspective, including the representations and roles of the major female figures.[38] Bradley provides a gynocentric alternative to the androcentric Arthurian tradition. Though she has the story set down by Morgaine, she places key women throughout—Igraine, Viviane, Morgause, Gwenhwyfar, Nimue, Niniane. At least one of these women is either present or using the Sight for each narrative incident, incidents told from their perspectives. Moreover, Bradley has readers see into the minds of these women, showing their reactions to issues that they face. The focus on these characters creates a balance with the major male characters—Gwydion/Arthur, Galahad/Lancelet, Gorlois, Uther, Taliesin, Kevin, Gwydion/Mordred. From Morgaine's first words, the novel establishes a balance between the interrelated binary oppositions of man | woman, Christian | pagan, Carleton/Camelot | Avalon. These binary oppositions form the central conflicts in the novel, and Bradley deftly deconstructs the opposition typically privileged in the Arthurian tradition—man, Christian, Carleton/Camelot.

Edel's tendency to excise aspects of Bradley's novel reverses her careful reconstruction of Arthuriana. The physical objects representative of Arthur's reign are perhaps the most glaring examples. Bradley's novel represents Excalibur as just one of several sacred items in existence in Celtic Britain: the Sacred Regalia. As part of Avalon's acceptance of Arthur as High King, Viviane shows him "the most sacred things in all your land," and we see these artifacts: a cup, platter, spear, and sword (Excalibur).[39] It is on this sword that Arthur must swear to deal fairly with Christian and pagan alike.

Bradley takes these items directly from the most influential medieval sources. Chrétien de Troyes, who introduced the Grail quest to the literary tradition, is the earliest source for most of them. Perceval receives a special sword and observes a bleeding lance, two gold candelabras, a grail, and a silver carving platter.[40] In Chrétien, the grail—which is undefined—is one of several holy relics. In Malory, we find similar holy objects, this time accompanied by angels: "two bare candils of wexe, and the thirde bare a towell, and the fourth a speare which bled mervaylously, that the droppis felle within a boxe which he hylde with hys othir hande."[41] Edel excises Bradley's Sacred Regalia—and by extension the holy relics presented in Chrétien and Malory—retaining only the phallic Excalibur.

Even more important for the androcentric miniseries, Edel omits Bradley's scabbard.[42] In Malory, after Arthur receives the sword and scabbard, Merlin tells Arthur that Excalibur's scabbard "ys worth ten of the swerde; for whyles ye have the scawberde uppon you ye shall lose no blood."[43] Key in all of this is the balance between sword and scabbard. The sword makes Arthur a valiant knight in attack; the scabbard protects him. Both Malory and Bradley present a seeming balance of masculine and feminine objects. Sword and spear are generally recognized as phallic symbols. They are weapons that penetrate. Scabbard and cup are acknowledged as feminine symbols. The sword goes into the scabbard; liquid goes into the cup. Bradley has maintained this balance in her novel. Viviane tasks Morgaine to make the scabbard, for "when that sword is carried into battle, it must be circled with all the magic we have."[44] As such, Morgaine is to "set into it every spell [she] know[s], that he who bears it into battle shall lose no blood."[45] Morgaine immerses herself in the task and creates the magical scabbard, the item that protects Arthur in battle. And it is the scabbard that, once it is clear that Arthur will not fulfill his oath to Avalon, Morgaine takes from him as he lies wounded at Glastonbury. She does not take Excalibur, the male symbol. Rather, she takes the scabbard, for "this at least she had a right to take—with her own hands she had fashioned it, the spells she had woven into it were her own."[46] Fleeing pursuers, Morgaine "whirl[s] it over her head, and fl[ings] it, with all her strength, far out into the Lake, where she s[ees] it sink into the deep and fathomless waters" so that "no human hand could ever reclaim it."[47] Morgaine's goal in this excursion was to take Excalibur, as the symbol of Arthur's oath to Avalon.[48] In casting the event in this way, though, Bradley has reinforced the masculine/feminine balance and opposition present in the novel. Morgaine is uncomfortable, or perhaps even unable, to reclaim Excalibur, Arthuriana's most notable phallic symbol. It has become part of the realm of men. The scabbard, however, was created in the realm of women, by a woman. In destroying the scabbard, Morgaine effectively withdraws Avalon's feminine protection from Arthur and his court, though leaving in place his masculinity. By eliminating this key feature of the relationship between Avalon and Arthur, Edel strips Bradley's narrative of its main mode of feminine protection. But he takes this a step further. In the miniseries, when Morgaine is unable to cross to Avalon, the mortally-wounded Arthur suggests that the Goddess needs an offering and hands Morgaine Excalibur. Morgaine hurls it over the lake's waters, above which it hovers, point down, and transforms into a glowing cross before disappearing, never having touched the water.[49] By presenting the phallic Excalibur as an offering to the Goddess, Edel erases the carefully crafted masculine/feminine balance and opposition, one of the central conflicts in Bradley's novel. The source of feminine power merges with and is replaced by the masculine, for by this point Avalon, source of feminine power, has already receded far into the mists and will eventually disappear.

This reandrocentrizing of the Arthurian narrative is just as apparent in how Edel represents the female characters. The women do appear in almost every scene in the miniseries; however, the changes entrench Bradley's characters into traditional gender roles. The alteration of Morgaine's character is pronounced.[50] Early in Edel's adaptation, Morgaine recalls that as a child she "was overjoyed that [her] father had come safely back to [them]." The scene behind the voiceover shows a young Morgaine joyfully running to her returning father.[51] The miniseries paints the pre-adolescent Morgaine as a dutiful, loving daughter, attributes absent from the novel. In fact, Bradley tells us flatly that "Morgaine was not a demonstrative child."[52] When Gorlois returns, Morgaine makes a brief appearance to greet him: "She [Igraine] brought Morgaine for a moment, washed and combed and pretty, for her to curtsey to her father, then had Isotta take her away to bed."[53] Bradley's Morgaine is not the adoring daughter that Edel presents.

Edel's infusion of emotionality becomes typical of Morgaine's portrayal in the miniseries. The Morgaine of Bradley's *Mists,* although not devoid of emotion, displays carefully-controlled and often delayed emotion. Repeatedly Morgaine comments upon her inability to weep or mourn. At Viviane's death—which occurs in the second half of the novel—Morgaine admits that the event aroused "weeping for only the second time that [she] could remember."[54] As an old woman, just before her return to Avalon, Morgaine is finally able to mourn: "And suddenly I began to cry. I wept, at last, for Accolon lying dead on his pall, and for Arthur who hated me now, and for Elaine who had been my friend ... and for Viviane, lying dead beneath a Christian tomb, and for Igraine, and for myself, for myself who had lived through all these things."[55] Once back in Avalon, Morgaine narrates that "It was a long season of mourning, and there were times when I wondered if I should mourn all my life and never again be free of it; but at last I could remember without weeping, and recall the days of love without unending sorrow welling up like tears from the very depths of my being."[56] Bradley presents a Morgaine who weeps very little in her life, until the very end. Edel changes Bradley's characterization, presenting a Morgaine who weeps repeatedly. She cries at her father's death,[57] at learning that Arthur was the King Stag,[58] when she confronts Viviane about her pregnancy,[59] when she visits Igraine at Glastonbury,[60] when Mordred kills Viviane,[61] and when Arthur dies.[62] In reversing Bradley's characterization and focusing the camera's gaze on Morgaine's weeping, Edel represents a character exhibiting reactions to trauma that the universal male viewer expects from a woman, placing Morgaine firmly within patriarchal gender roles.

Nowhere is the male gaze more obvious than in the realm that Mulvey first applied her concept of the gaze: erotic desire. As Mulvey says of Hitchcock, the "identification processes and liberal use of subjective camera from the point of view of the male protagonist draw the spectators deeply into his position, making them share his uneasy gaze."[63] As she argues throughout her work, the gaze embodies the viewer as masculine, and the miniseries exemplifies this operation. For instance, when Uther first sees Igraine, the camera focuses on Uther's face, his eyes intensely looking at her, then cuts to Igraine's face, assuming the perspective of Uther's gaze. The camera zooms to Igraine's eyes, directed up at the standing Uther, cuts to Uther's face, zooming to his eyes, then cuts to Gorlois's eyes, which move from Uther's to Igraine's. The camera follows Gorlois's gaze, focusing on Igraine, then moving to Uther, finally shifting back to Gorlois. In this brief scene early in the miniseries, the camera's gaze erases Bradley's gynocentric perspective. Although Igraine does look upon Uther with interest—though also with fear—the gaze from two men overpowers her, with Uther standing above her and Igraine's gaze fixed in an upward direction.[64] Moreover, Igraine is consistently portrayed as uncomfortable with the situation.[65]

The sexual scenes also reveal this masculine perspective. Bradley's Morgaine has sexual encounters with Arthur, Accolon, Lancelet, Kevin, and Raven, encounters that often define Morgaine's character. She is uneasy with Arthur yet loves him, and that experience remains with her until the end of her life. Her attraction to Lancelet, though not fully realized, haunts her throughout the novel. She has genuinely loving encounters with Accolon and Kevin, and she has brief, though fulfilling and repeated encounters with Raven. Edel has nearly eliminated desire from Morgaine. In the miniseries, we see Morgaine in precisely three moments of passion. The first is the King Stag episode with Arthur.[66] The novel, however, presents a second scene between Morgaine and Arthur, the morning after the Beltane rites. That encounter, Bradley stresses, is not between King Stag and Goddess, but between man and woman:

> This time in full awareness she could savor it, the softness and hardness, the strong young hands and the surprising gentleness behind his bold approach. She laughed in delight at the unexpected pleasure, fully open to him, sensing his enjoyment as her own. She had never been so happy in her life. Spent, they lay, limbs twined, caressing each other in a pleasant fatigue.[67]

It is a pleasant, joyful moment, whereas the previous evening had been a ritual, Morgaine not "in full awareness." It is a moment filled with pleasure for Morgaine, one that Edel chooses to omit. Instead, Edel only includes the ritualized sex between the King Stag and Virgin Huntress, both masked.[68]

These instances span the length of the miniseries. Morgaine's love for Accolon is but a footnote punctuated by a sexual encounter, again at Beltane, interspersed with cuts to the Arthur-Guinevere-Lancelot threesome, which relegates what in Bradley's novel was Morgaine's closest heterosexual relationship to a side-note, an interesting foil to the drunken encounter in Arthur's bedchamber.[69] The loving relationship of equals between Morgaine and Accolon is revealed in very few scenes, and it passes almost before it begins. The miniseries presents sexual attraction and tension between Morgaine and Lancelot, but that never comes to fruition, and Kevin the Harper, who provided Morgaine with more pleasure in the novel, is conspicuously absent.

More pressing, though, is Edel's deletion of Morgaine's brief though fulfilling erotic scenes with Raven. The first of these scenes is when Morgaine brings Nimue to Avalon for training. Morgaine and Raven make small cuts in their chests and lick the blood. After "Raven dr[aws] her [Morgaine] again into her arms," Morgaine thinks, "I gave up my maidenhood to the Horned One. I bore a child to the God. I burned with passion for Lancelet, and Accolon created me priestess anew in the plowed fields which the Spring Maiden had blessed. Yet never have I known what it was to be received simply in love."[70] Marilyn R. Farwell has described it as "an intensely erotic scene, and it is a crucial if not *the* crucial scene of the book,"[71] and the comparison between this experience and her previous sexual experiences cements its eroticism. 120 pages later, Bradley presents another erotic encounter between Morgaine and Raven. Raven has just revealed the theft of the Holy Regalia and regained her speech. She says, "I am afraid." Then,

> Morgaine lay down beside her, pulling Raven's cover over them both, and took Raven in her arms to still her shaking. As she lay quiet, listening to the other woman's breathing, she remembered the night she had brought Nimue there, and how Raven had come to her then, welcoming her to Avalon ... *why does it seem to me now that of all the love I have known, that is the truest* ... but she only held Raven gently, the other woman's head on her shoulder, soothing her.[72]

The scene's eroticism stems not from explicit details but, instead, from Morgaine remembering their previous encounter, from her thoughts of love, and from a slightly later sentence: "Morgaine lay silent, her fingers just entwined with Raven's."[73] Another, more explicit sexual scene occurs between Morgaine and Raven a few pages later. Frightened by the prospects of their mission, Raven pleads, "Morgaine, hold me, hold me, I am so frightened—."[74] The following paragraph, though brief and buried within the narrative, is important:

> Morgaine clasped her close and kissed her, rocking her like a child. Then, as if they entered together into a great silence, she held Raven against her, touching her, caressing her, their bodies clinging together in something like frenzy. Neither spoke, but Morgaine felt the world trembling in a strange and sacramental rhythm around them, in no light but the darkness of the dark side of the moon—woman to woman, affirming life in the shadow of death.[75]

It is an erotic scene, with the term *frenzy* emphasizing the sexual nature of it. To solidify the eroticism, Bradley again has Morgaine compare this to some of her past sexual experiences:

> As maiden and man in the light of the spring moon and the Beltane fires affirmed life in the running of spring and the rutting which would bring death in the field to him and death in childbearing to her; so in the shadow and darkness of the sacrificed god, in the dark moon, the priestesses of Avalon together called on the life of the Goddess and in the silence she answered them....[76]

Comparing this encounter with the Beltane ritual, the extended simile charges it sexually, making clear that what is happening is not two women simply seeking comfort. This is an erotic encounter, emphasized by the final sentences: "They lay at last quiet in each other's arms, and Raven's weeping was stilled at last. She lay like death, and Morgaine, feeling her heart slowing to stillness, thought, *I must let her go even into the shadow of death if that is the will of the Goddess*."[77] The experience propels Morgaine to the conclusion of the novel, providing her with, in a sense, peace of mind about her life's experiences and her mission to reclaim the Holy Regalia.[78] She has experienced multiple moments, fleeting and confined to her later life, of not only sexual but also spiritual closeness with another woman, moments that will influence the rest of her path in the story. And the scenes are in no way androcentric. These are not semi-pornographic scenes meant to titillate male erotic desire. Bradley has avoided erotic details, save Morgaine's emotional response.[79] This is not an instance of the male gaze. Farwell says of the first scene that

> symbolically she [Morgaine] moves from the female space of Avalon to and through the male space of Camelot, but the core of that female space—that which gives her power and strength—is sameness, is her connection with another woman. The space of movement in this novel is at its center defined by female desire, the desire of one woman for another, for Morgaine is at home only when she is in the arms of another woman.[80]

Edel's omission of these intensely passionate scenes is his final act of de-eroticizing Morgaine. On television Morgaine is not a sexual being, seemingly not capable of having a fulfilling sexual relationship with anyone. If the space of movement is, indeed, defined by female desire, then removing female desire, in essence, removes Bradley's space of movement, and as Farwell has noted, movement is central to the narrative.

In addition to these changes, the miniseries also expands the perspective beyond that of the women. In the novel, one of the female protagonists is present or witnesses via the Sight all narrated events. This is not the case in the miniseries, as Arthur's final battle illustrates. Bradley has the final battle occur during one of the Morgaine Speaks passages. She does witness some of the battle via the Well: "At first I saw only the wars raging up and down the land."[81] But in the next sentence, she indicates that the encounter between Arthur and Mordred is a mystery to her: "I never knew what came between Arthur and Gwydion."[82] She sees part of the final battle, sees Arthur and Gwydion searching for each other. She has a vision where she stands between them, attempting to stop the battle, but they attack, and "then I was in Avalon again, staring in horror at the mirror where I could see nothing, nothing but the widening stain of blood in the sacred waters of the Well."[83] She does not see the outcome, but instead physically goes to the battlefield, where she finds the two Gwydions dead and mortally wounded.[84]

The miniseries strips the scene of its mystery and female perspective. Very near the end, we see the final battle in all its masculine glory—swords hacking limbs, spears penetrating bodies—with none of the female protagonists present or using the Sight.[85] Grief over the funerals of Morgause and Viviane preoccupies Morgaine. The Sight comes to her when Arthur screams Mordred's name, for she sees parts of the battle through an orange filter. The filter clears—a cue that the Sight has ended—as Arthur and Mordred meet on the battlefield, with cuts to Morgaine desperately riding to the battlefield. She never witnesses the two fight, but the viewer sees the final battle between Arthur and Mordred in vivid detail. Morgaine sees only the end, where Arthur and Mordred, after looking into each other's pained eyes in what appears to be a father-son moment, deal each other their mortal wounds.[86] The camera's perspective in the miniseries is not entirely Morgaine's. And this shifting point of view, this shifting gaze, permeates the miniseries. Often Edel has female protagonists present to observe the events, but often they are simply not witnessing the action that Edel has filmed.

Bradley's medievalism focuses on the various Arthurian stories from the perspective of the major women in the medieval texts. She creates interrelated binary oppositions of male | female, Christian | pagan, Roman | Celtic as central sites of conflict, throughout casting these through a female gaze. In doing so, she draws upon research on the early Middle Ages, the medieval and post-medieval Arthurian tradition, and New Age mysticism. Bradley reverses the patriarchal, androcentric focus of nearly the entire Arthurian tradition, providing a view of the stories and characters from the female perspective, and thereby casting doubt on the place of these women in the traditional stories. Edel's medievalism, with its alterations and gaze, reverses Bradley's medievalism, reandrocentrizing the story, bringing it into line with not only much of Arthuriana but also much of television medievalism.

Edel's miniseries held the potential to be a shining exception to the androcentrism of 21st-century television medievalism. Although it does present Bradley's female characters as major elements, it fails to recreate her gynocentric perspective. That failure is precisely what some critics seem to have liked about it. Ken Tucker's *Entertainment Weekly* review reveals blatant sexism, calling Bradley's genre "fairy-tale feminist codswallop, swirled in purple prose" but admitting that he liked the miniseries and asserting that "Margulies embodies the sort of skirt-wearing power figure that Bradley and her legion of mostly female readers admire."[87] Clearly, Tucker values what Edel has done, and we can imagine that he would approve of the representations of women in such programs as *Merlin*, *Camelot*, *Legend of the Seeker*, and *The Vikings*, for Edel's Morgaine becomes the woman-warrior, inheriting and perpetuating the traits of this character type. The television medievalism present in these programs is in many ways the progeny of the TNT miniseries. Each of these shows had its debut long after Edel's *Mists of Avalon*. Each of them uses the male gaze as its point of view—through the protagonist, the scantily-clad warrior-woman, and the primary concern with the men in each storyline. One could speculate that a *Mists* miniseries that retained Bradley's gynocentrism may have had an effect on the medievalist television programs to come after it. That was obviously not the case. A miniseries conforming to Bradley's point of view, adopting the women's perspective, even a feminine gaze, could have initiated a trend in television medievalism that not only represented strong women but also provided those characters' perspective. The TNT miniseries, then, did very little to change the trajectory of later television medievalism. If "Bradley emphasizes that the power behind the throne is female, not male, and she keeps the viewpoint on women's magic, and Morgaine's role as a heroine, rather than men's battles,"[88] a gynocentric portrayal of Arthuriana, then Edel's effort is androcentrism masquerading as female perspective, an early harbinger of the proliferating trend of androcentric television medievalism.

Notes

1. Marion Zimmer Bradley, *The Mists of Avalon* (New York: Del Rey, 2001).
2. Gavin Scott, *The Mists of Avalon*, directed by Uli Edel, aired 15 July 2001 (Atlanta: TNT Original Video, 2012), DVD, hereafter referenced as Edel.
3. "Witchblade, The Mists of Avalon, Law & Order and NASCAR Cap Dramatic Summer for TNT," *Time Warner*, Press Release, 28 Aug. 2001, http://www.timewarner.com/newsroom/press-releases/2001/08/Witchblade_Mists_Avalon_Law_amp_Order_NASCAR_Cap_Dramatic_08–28-2001.php.
4. "Miniseries Review: 'The Mists of Avalon,'" Hollywoodwww, http://www.hollywood.com/static/miniseries-review-the-mists-of-avalon.
5. Tim Goodman, "Women Take Over Camelot: TNT's 'Mists of Avalon' Uses Female Touch To Improve Legend," *San Francisco Chronicle*, 13 July 2001, http://www.sfgate.com/entertainment/article/Women-take-over-Camelot-TNT-s-Mists-of-Avalon-2900158.php.
6. Ken Tucker, "Mists of Avalon," *Entertainment Weekly*, 13 July 2001, http://www.ew.com/ew/article/0,,255824,00.html.
7. Mary Magoulick, "Frustrating Female Heroism: Mixed Messages in *Xena*, *Nikita*, and *Buffy*," *The Journal of Popular Culture* 39 (2006): 731.
8. Magoulick, 730.
9. *Ibid*., 731.
10. *Ibid*., 743.
11. See Kathleen Kennedy, "Xena on the Cross," *Feminist Media Studies* 7 (2007), who sees *Xena* as offering a critique of patriarchy and violence, notes that the show "fails to offer us a roadmap to a more progressive future," 327.
12. Kennedy, 313.
13. Chris Chibnall and Steven Lightfoot, "Igraine," *Camelot*, Season 1, episode 8, directed by Michelle MacLaren, aired 20 May 2011.
14. Louise Fox, "Guinevere," *Camelot*, Season 1, episode 3, directed by Jeremy Podeswa, aired 8 April 2011.
15. Chris Chibnall, "Homecoming," *Camelot*, Season 1, episode 1, directed by Ciaran Donnelly, aired 23 February 2011.
16. Chibnall, "Igraine."
17. Magoulick, 731.
18. I should note here that *The Vikings* is a series on The History Channel, which at least alleges to present historically-accurate rather than mythical accounts.
19. Michael Hirst, "Rites of Passage," *The Vikings*, Season 1, episode 1, directed by Johan Renck, aired 3 March 2013.
20. Michael Hirst, "Trial," *The Vikings*, Season 1, episode 4, directed by Ciarán Donnelly, aired 24 March 2013.
21. Although I limit my comments here to television medievalism, film medievalism presents the same types of characters. One need only consider carefully Keira Knightley's Guinevere from

the 2004 *King Arthur*. In this rendering, Guinevere is a "Woad" (Pict) and a literal warrior woman. Typical of the character type, her dress reveals more skin than is covered, especially when in battle. In fact, *The Daily Mail* reported that "her breasts were digitally boosted for the advertising campaign." "My Flat Chest is a Turn-Off, Says Keira," *Mail Online, The Daily Mail*, http://www.dailymail.co.uk/tvshowbiz/article-395379/My-flat-chest-turn-says-Keira.html.

22. Leslie Workman, "Preface," *Studies in Medievalism* 8 (1996), quoted in Kathleen Verduin, "The Founding and the Founder: Medievalism and the Legacy of Leslie J. Workman," *Studies in Medievalism* 17 (2009): 20.

23. Tom Shippey, "Medievalisms and Why They Matter," *Studies in Medievalism* 17 (2009): 45.

24. Victoria Sharpe, "The Goddess Restored," *Journal of the Fantastic in the Arts* 9 (1998): 40.

25. Karen E. C. Fuog, "Imprisoned in the Phallic Oak: Marion Zimmer Bradley and Merlin's Seductress," *Quondam et Futurus* 1 (1991), 73.

26. Laura Mulvey, "Visual Pleasure and Narrative Cinema," in *Film Theory and Criticism: Introductory Readings*, ed. Leo Braudy and Marshall Cohen (New York: Oxford University Press, 1999), 838.

27. *Ibid.*, 843.

28. Mulvey, "Afterthoughts on 'Visual Pleasure and Narrative Cinema' inspired by King Vidor's *Duel in the Sun* (1946)," in *Feminist Film Theory: A Reader*, ed. Sue Thornham (New York: New York University Press, 1999), 122.

29. *Ibid.*, 125.

30. Mulvey later acknowledges a feminine gaze in film. Laura Mulvey, "Unmasking the Gaze," *Lectora* 7 (2001): 5–14.

31. Lois Tyson, "Feminist Criticism," in *Critical Theory Today: A User-Friendly Guide*, (New York: Routledge, 2006), 84–85.

32. For a more complete analysis of Morgan's character, see Charlotte Spivack, "Morgan le Fay: Goddess or Witch," in *Popular Arthurian Traditions*, ed. Sally K. Slocum (Bowling Green, OH: Bowling Green State University Popular Press, 1992) and Maureen Fries, "The Lady to the Tramp: The Decline of Morgan le Fay in Medieval Romance," *Arthuriana* 4.1 (1994): 1–18. See also Maureen Fries, "Female Heroes, Heroines and Counter-Heroes: Images of Women in Arthurian Tradition," in *Popular Arthurian Traditions*, ed. Sally K. Slocum (Bowling Green, OH: Bowling Green State University Popular Press, 1992): 5–17. For Morgan in popular culture, see Elizabeth S. Sklar, "Thoroughly Modern Morgan: Morgan le Fey in Twentieth-Century Popular Arthuriana," in *Popular Arthurian Traditions*, edited by Sally K. Slocum, (Bowling Green, OH: Bowling Green State University Popular Press, 1992): 24–35.

33. Of the women from whose perspective the story is told, only Igraine is not in some way Arthur's antagonist. Morgause, though not a direct antagonist, is mother to Mordred in the medieval sources. Nimue imprisons Merlin in Malory. Niniane—according to Norris J. Lacy and Geoffrey Ashe, *The Arthurian Handbook* (New York: Garland, 1998), 383—is a variation of Nimue. Guinevere's relationship with Lancelot initiates the downfall of Camelot in most medieval sources.

34. Bradley, xi.

35. *Ibid.*, x.

36. Spivak, 22, claims, "With Bradley, the role of Morgan le Fay thus comes full circle."

37. At the end of a number of Malory's sections we find devotional words. For instance, at the end of "The Book of Sir Tristram de Lyones" we find "here endith the secunde boke of Syr Tristram de Lyones, whyche drawyn was oute of Freynshe by Sir Thomas Malleorré, knyght, as Jesu be hys helpe. Amen." Sir Thomas Malory, *Works*, 2d ed., ed. Eugène Vinaver (Oxford: Oxford University Press, 1977), 511.

38. Fuog, 73, asserts that "the Arthurian legends are legends of a patriarchy, and have been retold by patriarchies for centuries."

39. Bradley, 203.

40. Chrétien de Troyes, *Arthurian Romances*, trans. William W. Kibler (London: Penguin, 1991), 419–21.

41. Malory, 603.

42. Lee Tobin, "Why Change the Arthur Story? Marion Zimmer Bradley's *The Mists of Avalon*," *Extrapolation* 34 (1993): 149–150, highlights Bradley's amplification of Malory's scabbard.

43. Malory, 36.
44. Bradley, 196.
45. *Ibid.*
46. *Ibid.*, 749.
47. *Ibid.*, 750.
48. *Ibid.*, 751.
49. Edel, "Home."
50. Here I focus on Morgaine because she is the primary protagonist; however, the same can be said for all of the women in the miniseries. Edel mutes Igraine's inner conflict, and she never renounces Christianity, as Bradley, 359–360, has her do. Morgause becomes not only what Maureen Fries calls a female counter-hero but also Felice Lifshitz's "destructive *dominae*" (Felice Lifshitz, "Destructive Dominae: Women and Vengeance in Medievalist Films," *Studies in Medievalism* 21 (2012): 161–190), and Edel completely omits Guinevere's internal conflict over her patriarchal gender role. For a detailed analysis of Bradley's treatment of Gwenhwyfar, see Tobin, 150–155. Edel also removes Nimue and Niniane from the miniseries.
51. Edel, "Marked Man."
52. Bradley, 78.
53. *Ibid.*, 85.
54. *Ibid.*, 502.
55. *Ibid.*, 757.
56. *Ibid.*, 758.
57. Edel, "Marked Man."
58. *Ibid.*, "Royal Proclamation."
59. *Ibid.*, ""Not Viviane's Pawn."
60. *Ibid.*, "Reunion."
61. *Ibid.*, "An Era in Eclipse."
62. *Ibid.*, "Another Incarnation."
63. Mulvey, "Visual Pleasure," 841.
64. Edel, "The Sign Foretold."
65. In the novel, the source of the Igraine/Uther connection is a shared previous life, just after Atlantis's destruction. This shared past life—eliminated in the miniseries—justifies in many ways the strong connection between the two and paves the way for a loving relationship.
66. Edel, "Beltane's Rites."
67. Bradley, 180.
68. This alteration also permits Edel to delay Arthur and Morgaine from identifying each other until very late in the film, whereas Bradley has the identification occur immediately after their morning encounter.
69. It is telling that the chapter title for this scene on the DVD is "Fertility Montage," stressing Guinevere's attempt to get pregnant over the passion between Morgaine and Accolon.
70. Bradley, 640.
71. Marilyn R. Farwell, "Heterosexual Plots and Lesbian Subtexts: Toward a Theory of Lesbian Narrative Space in Marion Zimmer Bradley's *The Mists of Avalon*," in *Arthurian Women: A Casebook*, ed. Thelma S. Fenster (New York: Garland, 1996), 320.
72. Bradley, 760.
73. *Ibid.*, 761.
74. *Ibid.*, 765.
75. *Ibid.*
76. *Ibid.*, 765–766.
77. *Ibid.*, 766.
78. Incidentally, this is also an instance of Morgaine being unable to weep, for the final sentence of this paragraph claims "And she could not even weep." *Ibid.*
79. The same is true of the threesome between Gwenhwyfar, Arthur, and Lancelet. Bradley, 449, ends the scene with "She pulled it [Morgaine's amulet] free and flung it into a corner, sinking back into her husband's arms and her lover's." Edel, however, shows just about as much as the FCC would permit.
80. Farwell, 329.

81. Bradley, 865.
82. *Ibid.*
83. *Ibid.*, 866.
84. *Ibid.*, 867.
85. Interestingly, these battle scenes are what some of the positive reviews stress. Goodman says the miniseries has "'Braveheart'-like battles and sexual escapades," and in his praise Tucker says that the miniseries is "regularly punctuated with impressively staged battle scenes between the 'Saxon barbarians' and the noble Christian/Goddess alliance."
86. Edel, "Casualties of Battle."
87. Tucker.
88. Tobin, 150.

Bibliography

Bradley, Marion Zimmer. *The Mists of Avalon*. New York: Del Rey, 2001.

Chibnall, Chris. "Homecoming." *Camelot*. Season 1, episode 1. Directed by Ciaran Donnelly. Aired 23 February 2011.

Chibnall, Chris, and Steven Lightfoot. "Igraine." *Camelot*. Season 1, episode 8. Directed by Michelle MacLaren. Aired 20 May 2011.

Chrétien de Troyes. *Arthurian Romances*. Trans. by William W. Kibler. London: Penguin, 1991.

Farwell, Marilyn R. "Heterosexual Plots and Lesbian Subtexts: Toward a Theory of Lesbian Narrative Space in Marion Zimmer Bradley's *The Mists of Avalon*." In *Arthurian Women: A Casebook*, ed. Thelma S. Fenster, 319–330. New York: Garland, 1996.

Fox, Louise. "Guinevere." *Camelot*. Season 1, episode 3. Directed by Jeremy Podeswa. Aired 8 April 2011.

Fries, Maureen. "Female Heroes, Heroines and Counter-Heroes: Images of Women in Arthurian Tradition." In *Popular Arthurian Traditions*, edited by Sally K. Slocum, 5–17. Bowling Green, OH: Bowling Green State University Popular Press, 1992.

_____. "The Lady to the Tramp: The Decline of Morgan le Fay in Medieval Romance." *Arthuriana* 4.1 (1994): 1–18.

Fuog, Karen E. C. "Imprisoned in the Phallic Oak: Marion Zimmer Bradley and Merlin's Seductress." *Quondam et Futurus* 1 (1991): 73–88.

Goodman, Tim. "Women Take Over Camelot: TNT's 'Mists of Avalon' Uses Female Touch To Improve Legend." *San Francisco Chronicle*. 13 July 2001. http://www.sfgate.com/entertainment/article/Women-take-over-Camelot-TNT-s-Mists-of-Avalon-2900158.php.

Hirst, Michael. "Rites of Passage." *The Vikings*. Season 1, episode 1. Directed by Johan Renck. Aired 3 March 2013.

Hirst, Michael. "Trial." *The Vikings*. Season 1, episode 4. Directed by Ciarán Donnelly. Aired 24 March 2013.

Kennedy, Kathleen. "Xena on the Cross." *Feminist Media Studies* 7 (2007): 313–332.

Lifshitz, Felice. "Destructive *Dominae*: Women and Vengeance in Medievalist Films." *Studies in Medievalism* 21 (2012): 161–190.

Magoulick, Mary. "Frustrating Female Heroism: Mixed Messages in *Xena*, *Nikita*, and *Buffy*." *The Journal of Popular Culture* 39 (2006): 729–755.

Malory, Sir Thomas. *Works*. Edited by Eugène Vinaver. 2nd ed. Oxford: Oxford University Press, 1977.

"Miniseries Review: 'The Mists of Avalon.'" Hollywoodwww. http://www.hollywood.com/static/miniseries-review-the-mists-of-avalon.

Mulvey, Laura. "Afterthoughts on 'Visual Pleasure and Narrative Cinema' inspired by King Vidor's *Duel in the Sun* (1946)." In *Feminist Film Theory: A Reader*, edited by Sue Thornham, 122–130. New York: New York University Press, 1999.

_____. "Unmasking the Gaze." *Lectora* 7 (2001): 5–14.

_____. "Visual Pleasure and Narrative Cinema." In *Film Theory and Criticism: Introductory Readings*, edited by Leo Braudy and Marshall Cohen, 833–44. New York: Oxford University Press, 1999.

"My Flat Chest is a Turn-Off, Says Keira." *Mail Online. The Daily Mail*. http://www.dailymail.co.uk/tvshowbiz/article-395379/My-flat-chest-turn-says-Keira.html.

Scott, Gavin. *The Mists of Avalon*. DVD Directed by Uli Edel. Aired 15 July 2001. Atlanta: TNT Original Video, 2012.

Sharpe, Victoria. "The Goddess Restored." *Journal of the Fantastic in the Arts* 9 (1998): 36–45.

Shippey, Tom. "Medievalisms and Why They Matter." *Studies in Medievalism* 17 (2009): 45–54.

Sklar, Elizabeth S. "Thoroughly Modern Morgan: Morgan le Fey in Twentieth-Century Popular Arthuriana." In *Popular Arthurian Traditions*, edited by Sally K. Slocum, 24–35. Bowling Green, OH: Bowling Green State University Popular Press, 1992.

Spivack, Charlotte. "Morgan le Fay: Goddess or Witch." In *Popular Arthurian Traditions*, edited by Sally K. Slocum, 18–25. Bowling Green, OH: Bowling Green State University Popular Press, 1992.

Tobin, Lee. "Why Change the Arthur Story? Marion Zimmer Bradley's *The Mists of Avalon*." *Extrapolation* 34 (1993): 147–157.

Tucker, Ken. "Mists of Avalon." *Entertainment Weekly*. 13 July 2001. http://www.ew.com/ew/article/0,,255824,00.html.

Tyson, Lois. "Feminist Criticism." In *Critical Theory Today: A User-Friendly Guide*, 83–133. New York: Routledge, 2006.

Verduin, Kathleen. "The Founding and the Founder: Medievalism and the Legacy of Leslie J. Workman." *Studies in Medievalism* 17 (2009): 1–27.

"Witchblade, The Mists of Avalon, Law & Order and NASCAR Cap Dramatic Summer for TNT." *Time Warner*. Press Release. 28 Aug. 2001. http://www.timewarner.com/newsroom/press-releases/2001/08/Witchblade_Mists_Avalon_Law_amp_Order_NASCAR_Cap_Dramatic_08-28-2001.php.

Gendering Morals, Magic and Medievalism in the BBC's *Merlin*

ELYSSE T. MEREDITH

Over five seasons (2008–2012), the BBC's *Merlin* distinctly gendered its depiction of magic and power. Set in a Camelot where magic and "the Old Religion" are outlawed by Uther Pendragon, the majority of male magic-users are portrayed positively, while female magic-users are almost always antagonists. This article examines the intersection of gender, magic, and morality within *Merlin* in relation to its literary antecedents and contemporary British concerns.

Merlin is a potent series to study, as it was broadcast on BBC One in the same primetime slot as the cultural monolith *Doctor Who*. Eventually broadcast in over 180 countries, the show's Saturday night slot pre-established it as fantastical "three-generational" family entertainment.[1] However, what began as a light-hearted, comedic action-adventure eventually became a dark and melancholic retelling. The illegality of magic drives the plot, which plays with the concept of the youthfulness of Arthurian knighthood by presenting it both metaphorically and literally: the story begins during Uther's reign, and the cast is composed primarily of twenty-somethings. Characters' traditional roles and backgrounds were modified by the producers in order to subvert the audience's expectations.[2] In this narrative, Arthur (Bradley James) is the legitimate son of Uther Pendragon (Anthony Head), who twenty years before outlawed magic and purged the country of all magic users; Morgana (Katie McGrath) is Uther's ward and illegitimate daughter; Merlin (Colin Morgan) is Arthur's servant; and Guinevere, called Gwen (Angel Coulby), is a blacksmith's daughter and Morgana's maid. While the series begins with its young protagonists balanced in gender and class, this intentional "subversion" was never meant to last.[3] The producers desired to "play games with the future legend" by presenting the world of Camelot (and, specifically, Morgana and Guinevere) as "the wrong way round" in order to create "intrigue" as to how the legend would be set to rights, with Morgana as villain and Gwen as queen.[4]

Emphasizing the illegality of magic and inverted character origins, this premise is completed by the introduction of the "Old Religion," which functions as a tripartite matriarchal system of magic, ethics, and belief that is opposed to the ethical and social structures of secular and patriarchal Camelot. This overarching conflict between two ethical/social systems, enacted in the personal spheres and moral lives of the primary characters, results in an androcentric narrative that vilifies women with power. By breaking down the class and gender equity established by its inverted character origins, *Merlin* develops a conservative and patriarchal rendering of a "golden age" that is viewed as superior to its more egalitarian beginnings. Viewed as a whole, *Merlin* uses its Arthurian narrative to depict an uneasy concern with contemporary social advancements, creating a narrative of justified inequality.

To explore this concept fully, this article examines the Old Religion to elaborate on the separate functions of destiny and fate. Reading the valences of destiny, fate, and the Old Religion against the social ethics and personal moralities developed by the narrative demonstrates how the series genders the use of magic. The characters Morgana, Mordred, Merlin, and Guinevere have particularly complex interactions with power, gender, and magic and deserve special attention. Overall, *Merlin* is a medievalism of gendered morality that comments on religion, monarchy, and the role of governmental devolution in contemporary British society. This article demonstrates how contemporary Arthuriana can be potently nationalistic, serving as quasi-propagandistic material that expresses anxieties over contemporaneous political and social issues by depicting corresponding subjects as inherently dangerous.

The Old Religion

The series' depiction of the Old Religion is strongly matriarchal and inspired by Celtic-based strains of neo-paganism. Its most obvious influence is Marion Zimmer Bradley's 1983 novel *The Mists of Avalon*, which introduced the concept of a quasi–Celtic matriarchal neo-pagan religion to Arthuriana.[5] Like the Morgana of *Merlin*, Bradley was a priestess of the Triple Goddess, a formulated deity found in Robert Graves' 1948 text *The White Goddess* (who also conceived of the titular deity).[6] The term "Old Religion" is itself drawn from Charles G. Leland's 1899 text *Aradia, or the Gospel of Witches*.[7] While there is little doubt that these neo-pagan concepts are "contemporary creations" from "patently contrived or even erroneous" sources, their conscious creation of an archaic lineage suits *Merlin*'s fantastical Arthurian setting.[8]

The series establishes that "the Old Religion is the magic of the earth itself ... the essence which binds all things together" as well as "the source of [Merlin's] power."[9] Every magic in the show is derived from the Old Religion, and the Old Religion is shown to be "very real."[10] Even Merlin's mentor Gaius (Richard Wilson) was initially a practitioner of the Old Religion, though he rejected both the Old Religion and magic in favor of science. While in his magical practices Merlin acts as if his magic is separate from the Old Religion, he taps into the Old Religion when he accesses the power of the White Goddess (depicted as a shining light) in the fifth season to undo Morgana's brainwashing of Guinevere.[11] However, the end of the series alters the position of the Old Religion within the narrative's universe, asserting in the two-part finale that "*magic* is the fabric of this world" from which Merlin was born, not the Old Religion.[12]

This shifted emphasis from Old Religion to magic is best explored by examining the Old Religion's relationship with Camelot and the distribution of power and gender in its practitioners. There are two primary groups of practitioners of the Old Religion in *Merlin*. The first, the druids, are generally patriarchal and peaceful; when they commit violent acts, the series emphasizes that these are a response to Uther's violent oppression.[13] The druids are a separate entity from the organized Old Religion, whose priesthood is limited to women (upon Morgause's death Morgana becomes the last high priestess). Male practitioners were excluded from leadership; male priests, though extant, were subordinate.[14]

The organized Old Religion is presented as violent and malevolent, with priestesses using magic for torture, brainwashing, forced conversion, and blood sacrifices.[15] It is a religion without hope and comfort, focused on matriarchal power and the balance between life and death.[16] Indeed, Merlin outright condemns a high priestess of the Old Religion, proclaiming that she practices a "selfish and cruel magic."[17] Yet while all high priestesses act as enemies of the protagonists, every male practitioner of the matriarchal Old Religion is depicted as firmly rejecting at least one high priestess in order to join Merlin's side. By the end of the series, "follower of the Old Religion" is synonymous with "evil sorceress."[18]

While the Old Religion is depicted as sinister, it is also presented as undeniably real, and its trappings act as a cultural background for Camelot. While Uther banned the practice of magic and the Old Religion from Camelot, he allows the celebration of Beltane, Samhain, and Ostara. These high holy days in the Old Religion form the foundation of the cultural year of Camelot, and are also actual holidays celebrated in some neo-pagan traditions, particularly Celtic neo-paganism.[19] This casual inclusion of the Old Religion within a primarily secular Camelot reflects the role of Christianity within earlier Arthuriana, particularly medieval texts such as the works of Chrétien de Troyes. While the influence of Christianity on society is palpable in these texts, it is felt primarily as part of the structures of everyday life, becoming a significant force only when introduced as an external compulsion, as with the grail quest in the Prose *Lancelot*. Similarly, while the cultural trappings of the Old Religion in *Merlin* are kept by Camelot, the Old Religion itself is only a force within Camelot when introduced by an antagonist.

While the depiction of the Old Religion draws upon previous representations of religion in Arthuriana (including contemporary neo-paganism and medieval Christianity), it also creates a religious atmosphere that resonates with the state of Christianity within contemporary British culture. Christianity "has had an irreversible effect on the shaping of time and space in this part of the world," and like the Old Religion, the Church of England occupies a unique space in both society and state.[20] While the Church of England retains a measure of political influence, an increasingly secularized British society enjoys the cultural trappings of holidays but generally ignores or rejects the religion's theological and spiritual sides as outdated or (in some cases) morally bankrupt.[21]

In both contemporary Britain and *Merlin*'s Camelot the secularization of society is a recent event, though in *Merlin* it is a radical secularization signaled not simply by withdrawal of state support and "the cessation of all interest in religious perspective, practice, and institutionalized features" but by the violent Great Purge.[22] Merlin's use of magic (practiced outside the Old Religion) reflects how "the majority of British people ... persist in believing ... but see no need to participate with even minimal regularity in their religious institutions"; that is, "believing without belonging."[23] By positioning the Old Religion in Camelot as culturally analogous to Christianity within Britain, the series allows itself to explore contemporary concerns with religion through the characters' interactions with the Old Religion, its magic, and its practitioners.

Destiny, Fate, Ethics and Morality

In addition to the Old Religion, two other concepts shape the moral choices of the characters in *Merlin*: fate and destiny. These are presented as distinct from the Old Religion as well as from one another; Merlin himself states in the final season that "there is a difference between fate and destiny."[24] Within the narrative's universe, fate is inevitable, while destiny is an imperative, a requirement of an individual. This is reinforced by the opening sequence of each episode, which reiterates that the "destiny of a great kingdom rests on the shoulders of ... Merlin." This destiny, which is to aid Arthur in "unit[ing] the land of Albion," is Merlin's driving force.[25]

As magic is illegal in Camelot, Merlin's use of magic necessarily places him outside Camelot's ethical structures, and his realization of his destiny becomes his moral imperative. He must act in its favor regardless of whether the act is ethically dubious, as his destiny overrides the ethics of Camelot. Thus, in the course of the series Merlin regularly deceives Camelot's court and his friends, attempts to kill Morgana and Mordred, and kills several others.[26] From his perspective, these are inevitable necessities required by his destiny. His choices are legitimized by his second mentor the Great Dragon (John Hurt), who states in the first episode that "there is no right or wrong, only what is or isn't."[27] The Great Dragon urges overtly malevolent actions (such as killing Uther, Morgana, and Mordred) as not only acceptable but necessary because they stop destined evil. By combining destiny, necessity, and a sense of righteousness, the Great Dragon influences Merlin to view the consequences as justification for his deeds.

Because destiny is dependent on correct action and choice in this universe, it is possible for destiny to be unrealized; as Gaius states, "the future has many paths."[28] Conversely, fate is inevitable and, in this universe, determined by literary precedent. Indeed, at the end of the series Gaius says that "there's some parts that are woven so deeply into the fabric of the world that nothing can be done to change them."[29] Even in the first season, *Merlin* is fatalistic about Arthur's eventual doom, entitling the episode introducing Mordred "The Beginning of the End."[30] The "fate" of *Merlin* is that the greater Arthurian mythos must eventually reassert itself: Morgana will become an antagonist, Arthur will establish the Round Table, Guinevere will become queen, Mordred will mortally wound Arthur, and Arthur will become the "Once and Future King."[31] Although this all occurs, Albion is only partially united when Arthur is killed by Mordred. At the end of the series, Merlin's destiny is not fulfilled, but fate (literary precedent) has reasserted itself despite the series' subverted beginnings.

While there are multiple competing ethical systems in *Merlin* (those of the Old Religion and Camelot, where magic is banned), there seems to be only one overarching moral system within the universe. Combined with the reality of the malevolent Old Religion, the Great Dragon's denial of right and wrong suggests that the world of *Merlin* is naturally nihilistic and fatalistic. Morality is therefore a structure of dispassionate social ethics based in consequentialism, where consequences justify the deed. There is no right or wrong; one must simply attempt to fulfill one's destiny and accept one's fate.

Gendering Magic

The intersection of gender, magic, and morality within *Merlin* evokes both medieval misogynies and neo-pagan principles. While neo-paganism "values, indeed divinizes, the *active self*," medieval conceptions of gender based on Aristotle framed males as active and females as passive; to have women in active roles would be a perversion.[32] In *Merlin*, female power is typified as both active and mystical, and presented as especially dangerous to men: women with magic often exercise control over or supplant men, as with Morgana's quest to usurp Uther and Arthur. Whereas from the very first episode unredeemable villains tend to be female, antagonists with nuanced motivations tend to be male. Male characters often commit ambiguous or outright malevolent deeds, yet are exonerated through good consequences or repentance. Furthermore, the majority of recurring male antagonists act as servants for female sorcerers. Their servitude is typified by a fascination or enthrallment with the sorceress in question, often with sexual overtones.

This gendered depiction of villainy is particularly prominent because the majority of female adversaries are magic-users while the majority of male enemies are not. Furthermore, female characters are generally either evil or controlled by evil (usually generated by other women); those few women who are not evil or possessed by it end up dead. By the series finale, only one living female character has not committed an evil act: Merlin's mother, who appears insignificantly in four episodes, and does not possess magic.[33] Only one female character, Finna, uses magic solely for good.[34] An initiate in the Old Religion, Finna appears once, where she is shown to be subservient to her male mentor and Merlin. She is so reverential to Merlin that she kneels in front of him, and at the episode's end she commits suicide in order to protect him.

The disparity in treatment of antagonists based on gender is visible in the overall treatment of gender, villainy, and magic within *Merlin*. While the total numbers of male and female antagonists are roughly equivalent, there are over thirty more female adversaries with magical powers than male adversaries with magical powers.[35] In the first and second seasons the number of male and female villains with magic is roughly equal, but with Morgana's transformation into an antagonist in the third season the female enemies with magic (counted per appearance) grossly outnumber the men. In contrast, over the course of the series there are a minimum of fifty-six different instances of male heroism, while there are less than ten instances of female heroism. However, while the numerical disparity demonstrates the relationship of the narrative's constructed moral universe with this gendered depiction of magic, the portrayal of gendered morality and magic is most clearly seen in the treatment of four particular characters:

the antagonists Morgana and Mordred, and the protagonists Merlin and Guinevere.

The Danger of a Powerful Morgana

The series' critique of active female power is elucidated by the progression of Morgana's character. Though initially a protagonist, her devolvement into antagonism was always planned, with a direct correlation between the growth of her magical powers and villainy.[36] The narrative intentionally constructs parallels between Merlin and Morgana with the Great Dragon's declaration that Morgana "is the darkness to [Merlin's] light, the hatred to [his] love."[37] Yet as both characters seek the end of Uther's tyranny and the return of magic, their primary difference is that Morgana is an antagonist and Merlin the hero.[38]

Like the Morgana of medieval Arthuriana, *Merlin*'s Morgana is a noblewoman "with the leisure to devote [herself] to the study of magic," and thus positioned "at an interesting tangent to the courtly world, challenging or unsettling its norms."[39] However, in her earliest literary appearances Morgana was not an enemy; late medieval narratives presented her as Arthur's caretaker, healer, and companion.[40] *Merlin* discards this complicated relationship. Yet the series' creators intended Morgana to be sympathetic, embracing the "dark side through circumstance."[41] Notably, the series presents Merlin multiple opportunities to offer magical friendship to Morgana, but he rejects every opportunity and isolates her from help. Indeed, he attempts to kill her three times.[42] The first of these is a shocking betrayal that confirms Morgana on her path to antagonism. Living under the constant threat of emotional, physical, and verbal abuse, Morgana finally embraces her power to fight Uther at the encouragement of her half-sister, Morgause.

The vilification of Morgana is potent even when Morgana is still ostensibly a protagonist. Even in the first season the Great Dragon is utterly derogatory of her, encouraging Merlin to view her as a rival.[43] Similarly, Gaius's outright fear of Morgana's power is remarkable because he is Morgana's friend, mentor, and physician.[44] Obliquely suggesting that Morgana's powers are evil, Gaius uses her status as Uther's ward as an excuse to isolate her.[45] This paternalistic protectionism leaves Morgana vulnerable to discovery and Morgause's influence.

Morgana's conversion to a sorceress is depicted as understandable, sympathetic, and relatable, stemming from oppression and abuse, but once inducted into the Old Religion she becomes unredeemable, a willing usurper and villain. This is accomplished by a visual cue that the series' creators termed a "witchiepoo moment."[46] This infantilizing diminutive refers to a villainess smirking toward the camera, suggesting a relish of villainy and depraved motivation.

While this visual cue is employed by other female antagonists, it becomes Morgana's signature expression, the literal face of her desire for the throne.

The Absolution of Mordred

While the narrative transforms Morgana from sympathetic protagonist to single-minded villain, Mordred's path is a rough reversal of Morgana's. Initially, Mordred's introduction disrupts the associations between villainy, magic, and gender. The Great Dragon primes the audience to fear the child-sorcerer Mordred (Asa Butterfield), informing Merlin of ancient prophecies that "speak of Morgana and Mordred united in evil" and Mordred's destiny to kill Arthur.[47] This is particularly disturbing because Mordred is indebted to Arthur, who rescued him from Uther. Yet when Mordred reappears as a grown man (Alexander Vlahos) in the fifth season, he condemns and attacks Morgana, saves Arthur, and becomes a knight of the Round Table.[48] This is a startling development, for in his first appearance Mordred forms a strong bond with Morgana.[49] This bond was possibly also inspired by Marion Zimmer Bradley, who was the first to cast Morgana as Mordred's mother; in medieval texts, Morgana is at best his aunt, if not completely unrelated.[50] Contrasted with his earlier portrayal in *Merlin*, Mordred's gentility and kindness in protecting Arthur only emphasizes that Morgana is beyond redemption.

Mordred's supposedly innate evil is completely elided in the fifth season. His betrayal of Arthur results not from inherent malevolence but from the death of his childhood sweetheart. Executed for the attempted assassination of Arthur, she gives Mordred a plausible impetus to return to Morgana.[51] However, at Morgana's side Mordred is no longer portrayed as a dangerous sorcerer. Rather, he is the tool Morgana uses to kill Arthur. Instead of a powerful union, Morgana and Mordred's relationship retreads the show's earlier positioning of male antagonists as subservient to sorceresses.

Good Guinevere

The danger of powerful women is marked in the portrayal of Guinevere, called Gwen. The most prominent and lauded of women in *Merlin*, she is the only continuously positive female character. Wise and forgiving, she cares for both family and strangers; she even tends an ill Uther, who executed her father and rejected her as a suitable wife for Arthur due to her class.[52] Additionally, she is an impetus and reward for male heroism, literally described as a "damsel in distress."[53] Once she becomes queen, Gwen wields legitimate but passive power, her orders being enacted by others. In commanding the knights of the

Round Table, she becomes a motivation and inspiration that does not challenge Camelot's masculine society.

This positive portrayal replicates medieval Arthuriana, where "even in the harsh judgment of the Post-Vulgate, there appears to be a persistent desire to give the queen a positive profile or at least to detect favorable traits in her character."[54] Like *Merlin*'s Gwen, the medieval Guinevere was associated with "eloquence, courage, humour, [and the] power to instruct" as well as "kindness, compassion, [and] saintliness."[55] In particular, English romances tended to avoid blaming Guinevere, who was "a popular figure of identification" for the English "in defiance of the continental adversaries and their contribution" to the guilty Guinevere tradition.[56] It is consequently crucial that Gwen's negative actions—including her affair with Lancelot, betrayal of Arthur, and murder—occur under the influence of Morgana's magic.[57] Gaius states outright that Gwen is still "innocent and perfect"— "the only evil in her is Morgana's."[58] The absolved Gwen is ethically pure, an acceptable leader for Camelot, and *Merlin* continues the English tradition of Guinevere as "the realization of national character."[59]

However, Gwen's virtue is insufficient to redeem Morgana, with whom she initially enjoys a strong friendship.[60] This fails when Morgana prophetically dreams that Guinevere will be crowned queen of Camelot.[61] This dream causes Morgana to transfer her hatred of Uther (as his illegitimate and abused child) onto Guinevere (her former maidservant becoming her usurper).[62] While this supposedly highlights the changing social equity in Arthur's Camelot (where even a serving girl can become queen), it is Gwen's inherent integrity that is critical in transforming Arthur into the egalitarian king of the Round Table, not her class.[63]

Her political power is gained through marriage; conversely, Morgana rejects male rule and actively pursues alternative modes of power. By distributing activity/rebellion/evil and passivity/tradition/good to Morgana and Guinevere respectively, the series reiterates medieval misogynies. By withdrawing the positive traits Morgana bears in medieval narratives and contemporary feminist retellings, *Merlin* transforms Morgana's response to Uther's oppression into a thing to ridicule; by divesting Guinevere of negative aspects, *Merlin* takes the nationalistically flattering Guinevere found in medieval English romances to an extreme.

Contemporary Concerns

The significance of crowning a pure Guinevere is crystallized in the series' two-part finale, which includes two unexpected references to British unionism. In the penultimate episode, Arthur gives a final speech to the knights of the

Round Table where he reminds them that they fight for "the future of the united kingdoms."[64] The penultimate scene of the entire series places Gwen alone on the throne, with the ringing of "long live the queen" serving as the last words of the series.[65]

While it is logical for Gwen to become sole ruler of Camelot upon Arthur's death (the final season often shows her ruling wisely in his stead), this choice makes the final political situation of Camelot mirror the modern United Kingdom, with Gwen paralleling Queen Elizabeth II. This extension of Guinevere as "the realization of national character" creates an almost propagandistic conclusion that invites and even demands reassessment of its medievalism in relation to contemporary British concerns.[66]

The most explicit contemporary references are located in the Old Religion. Its roots are founded in 20th-century Celtic neo-paganism, but its depiction is culturally analogous to the role of Christianity in contemporary Britain. As the narrative asks us to hope for the golden age of Arthur's secular Camelot, it suggests that even a "very real" religion is detrimental to society. By basing the malevolent Old Religion on Celtic neo-paganism, the series also impugns a minority religion in Britain and reasserts a historical fear of and discrimination against the Celtic Other. This is particularly relevant as both Morgana and Merlin are played by Irish actors (Colin Morgan and Katie McGrath). While Merlin uses an English accent, Morgana uses her actress's native accent. Morgana's villainy is in part defined by her priestesshood in the Old Religion, and the use of McGrath's native accent recalls that the Irish have historically "been represented ... as a religious threat."[67] The distinction between the series' primary two magic users, then, becomes enshrined in gender, nationality, and religion, aurally cuing benevolence or malevolence.

In countering Morgana's violent revolution and the Celtic Old Religion with secular unionism, *Merlin* unites anxieties over independent female power and the Celtic Other, and suggests a broader concern with the devolution of the Scottish, Welsh, and Northern Irish governments as well as the Scottish independence referendum of 2014.[68] As these historically Celtic nations are no longer treated as "peripheral appendages to England, but as national entities with clear institutional presence," it is clear within the political United Kingdom that "England is not Britain."[69] Equally, in *Merlin* Camelot is not Albion; Arthur's destiny to unite Albion and return in its hour of need may reveal an "anxiety" about the potential "end of the United Kingdom" that devolution could cause (most prominently indicated by the Scottish referendum).[70] Where the British Empire once "reconciled the Celtic periphery to an unequal union (or semi-union) with England," Arthur's Camelot attempts the same for Albion.[71] Yet the series does not end with the legendary Arthur ruling an empire; neither does it end with Morgana's revolution nor Merlin's golden age of magic. Instead, it leaves us ruled by the gentle, passive, wise, and secular Guinevere, who recalls the current British Queen.

Conclusion: The Intersection of Gender, Magic and Morality

Throughout *Merlin*, whether power is political or magical, when wielded by a man it is commonly held as acceptable; if he performs evil, there is often a woman controlling him. If magical power is wielded by a woman, it is generally religious, mystical, and evil. Though shown to be real, the matriarchal quasi–Celtic Old Religion is also vilified and maligned, linked to Morgana as both the source of her power and an intrinsic part of her unambiguous villainy. Yet when its holidays are celebrated in the homosocial, patriarchal Camelot, they are positive secular events.

As *Merlin* devolves from a light-hearted adventure of the literal youth of Camelot into a melancholy reiteration of the end-matter of Arthurian legend, its positive and negative representations of power, religion, and secular authority grow increasingly polarized along gender lines. Drawing upon medieval antecedents as well as more recent retellings, it reworks these conceptions in a manner that implies contemporary anxieties concerning women, religion, politics, and the future of the United Kingdom. As a cultural artifact, *Merlin* suggests a deep-seated unease about the role of women and religion in politics and the United Kingdom. In its narrative, only when women and religion are passive, pure, and/or minimally included are they tolerable and/or a source of social improvement.[72] Depicting Morgana's rejection of Uther's oppressive anti-magic policies as a violent campaign belies social anxieties concerning women, social power, religion, and the Celtic Other; that the series ends without magical freedom implies that stagnation is better than revolution. Its contemporary concerns and mythical setting indicate that little has changed since the age of legends; the only difference is that magic has left the world.

In the final episode of the series, Merlin sends Arthur's body to Avalon. With "long live the queen" still ringing in the audience's ears, the final scene depicts an elderly Merlin walking past a present-day Avalon that looks suspiciously like Glastonbury Tor. Merlin still waits for the return of Arthur: a male ruler, a political unifier and conqueror, a secular authority, and a legendary hero. He waits for a figure who will resolve the series' social concerns and "bring about the world we dream of."[73] As the fatalistic universe of *Merlin* draws its curtain, this world has not come, and the audience must wait with Merlin.

Notes

1. *Merlin: The Complete First Season*, Commentary for "The Dragon's Call" with producers Julian Murphy and Johnny Capps, DVD (United Kingdom: Shine Limited and BBC Wales, 2010); *Merlin: The Complete First Season*, Commentary for "Lancelot" with Johnny Capps and Ed Fraiman, DVD (United Kingdom: Shine Limited and BBC Wales, 2010); *Merlin: The Complete First Season*, "Behind the Magic Part One," DVD, produced by Mark Procter and Zoë Rushton (United Kingdom: BBC Wales, 2010); *The Adventures of Merlin: The Complete Second Season*, "Cast and Crew Introduction to Merlin Season 2," DVD (United Kingdom: BBC Wales, 2011); *The Adventures of Merlin: The Complete Third Season*, "Merlin Season 3 Comic-Con San Diego 2010," DVD (United Kingdom: Shine Limited and BBC Wales, 2012); *The Adventures of Merlin: The Complete Fourth Season*, "Making of Merlin *Season 4*," DVD (United Kingdom: Shine Limited and BBC Wales, 2013).

2. *First*, Commentary for "The Dragon's Call"; *Merlin: The Complete First Season*, "Lancelot," BBC One, October 18, 2008, written by Jake Michie, directed by Ed Fraiman; *First*, Commentary for "Lancelot"; *First*, "Behind the Magic Part One."

3. *First*, Commentary for "The Dragon's Call"; *First*, Commentary for "Lancelot"; *The Adventures of Merlin: The Complete Fourth Season*, Commentary for "The Wicked Day" with Colin Morgan and Alice Troughton (United Kingdom: Shine Limited and BBC Wales, 2013); *The Adventures of Merlin: The Complete Second Season*, "The Making of Merlin," DVD, directed by Mark Procter, produced by Gillane Seaborne (United Kingdom: BBC Wales, 2011); *Third*, "Comic-Con 2010."

4. *First*, Commentary for "The Dragon's Call."

5. Carolyne Larrington, *King Arthur's Enchantresses: Morgan and Her Sisters in Arthurian Tradition* (New York: I.B. Tauris, 2006), 189–192; Diana L. Paxson, "Marion Zimmer Bradley and *The Mists of Avalon*," *Arthuriana* 9 (1999): 120. Bradley's work is not the only contemporary Arthurian literature that *Merlin* has drawn upon; two of Mary Stewart's four Arthurian novels give their names to the episodes *The Crystal Cave* and *The Wicked Day* (Larrington, 179).

6. Robert Graves, *The White Goddess: A Historical Grammar of Poetic Myth* (New York: Farrar, Straus and Giroux, 1948), amended and enlarged edition 1966, 64; *The Adventures of Merlin: The Complete Fourth Season*, "The Secret Sharer," BBC One, November 12, 2011, written by Julian Jones, directed by Justin Molotnikov; Larrington, 193–194; Paxson, 114; Michael York, "Invented Culture/Invented Religion: The Fictional Origins of Contemporary Paganism," *Nova Religio: The Journal of Alternative and Emergent Religions* 3 (1999): 138.

7. York, "Invented Culture," 138.

8. *Ibid.*

9. *Merlin: The Complete First Season*, "Le Morte d'Arthur," BBC One, December 13, 2008, written by Julian Jones, directed by David Moore.

10. *First*, "Le Morte d'Arthur."

11. *The Adventures of Merlin: The Complete Fifth Season*, "With All My Heart," BBC One, December 1, 2012, written by Richard McBrien, directed by Alice Troughton.

12. *First*, "Le Morte d'Arthur"; *The Adventures of Merlin: The Complete Fifth Season*, "The Diamond of the Day, Part One," BBC One, December 22, 2012, written by Jake Michie, directed by Justin Molotnikov (qtd., my emphasis).

13. *The Adventures of Merlin: The Complete Third Season*, "The Crystal Cave," BBC One, October 9, 2010, written by Julian Jones, directed by Alice Troughton; *The Adventures of Merlin: The Complete Fourth Season*, "A Herald of the New Age," BBC One, December 3, 2011, written by Howard Overman, directed by Jeremy Webb; *The Adventures of Merlin: The Complete Fifth Season*, "The Drawing of the Dark," BBC One, December 15, 2012, written by Julian Jones, directed by Declan O'Dwyer.

14. *Fourth*, "The Secret Sharer"; *The Adventures of Merlin: The Complete Fifth Season*, "The Kindness of Strangers," BBC One, December 8, 2012, written by Richard McBrien, directed by Declan O'Dwyer.

15. *The Adventures of Merlin: The Complete Third Season*, "Love in the Time of Dragons," BBC One, November 6, 2010, written by Jake Michie, directed by Alice Troughton; *The Adventures of Merlin: The Complete Fourth Season*, "The Darkest Hour, Part One," BBC One, October 1, 2011, written by Julian Jones, directed by Alice Troughton; *The Adventures of Merlin: The Complete*

Fourth Season, "A Servant of Two Masters," BBC One, November 5, 2011, written by Lucy Watkins, directed by Alex Pillai; *Fifth*, "With All My Heart."

16. *Merlin: The Complete First Season*, "Excalibur," BBC One, November 15, 2008, written by Julian Jones, directed by Jeremy Webb; *First*, "Le Morte d'Arthur."

17. *First*, "Le Morte d'Arthur."

18. *Fifth*, "The Kindness of Strangers."

19. *Merlin: The Complete First Season*, "The Dragon's Call," BBC One, September 20, 2008, written by Julian Jones, directed by James Hawes (qtd.); *Merlin: The Complete First Season*, "The Poisoned Chalice," BBC One, October 11, 2008, written by Ben Vanstone, directed by Ed Fraiman; *First*, "Le Morte d'Arthur" (qtd.); *Fourth*, "The Darkest Hour, Part One" and "The Secret Sharer"; *The Adventures of Merlin: The Complete Fourth Season*, "The Hunter's Heart," BBC One, December 10, 2011, written by Richard McBrien, directed by Jeremy Webb; *The Adventures of Merlin: The Complete Fourth Season*, "The Sword in the Stone, Part One," BBC One, December 17, 2011, written by Jake Michie, directed by Alice Troughton. In the absence of a new religion, the series introduces the character of Geoffrey of Monmouth as a priest-like figure within Camelot. He is depicted as the court genealogist and librarian, who is occasionally called upon to perform wedding ceremonies and crownings. This causes an author of the Arthurian corpus to become an authority within the narrative.

20. Grace Davie, "Religion in Europe in the 21st Century: The Factors to Take into Account," *European Journal of Sociology* 47 (2006): 273.

21. Peter W. Edge, "Secularism and Establishment in the United Kingdom," in *Religion, Rights and Secular Society: European Perspectives*, ed. Peter Cumper and Tom Lewis (Northampton, MA: Edward Elgar, 2012); 42 and 44.

22. Callum G. Brown, *The Death of Christian Britain: Understanding Secularisation 1800–2000* (New York: Routledge, 2001), 30; Michael York, "New Age Commodification and Appropriation of Spirituality," *Journal of Contemporary Religion*, 16 (2001): 362 (qtd.). The Great Purge was compared to the Holocaust in *Merlin: The Complete First Season*, Commentary for "The Beginning of the End," with Angel Coulby, Colin Morgan, Katie McGrath, and director Jeremy Webb, DVD (United Kingdom: Shine Limited and BBC Wales, 2010), but there is also a passing resemblance to the Dissolution of the Monasteries. This could suggest a reading whereby Uther is equivalent to Henry VIII, Arthur to Edward VI, magical Morgana to the Catholic Mary I, and Guinevere to Elizabeth I.

23. Grace Davie, *Religion in Britain since 1945: Believing without Belonging* (Cambridge: Blackwell, 1994), 2 and 74–75.

24. *The Adventures of Merlin: The Complete Fifth Season*, "The Disir," BBC One, November 3, 2012, written by Richard McBrien, directed by Ashley Way.

25. *First*, "The Dragon's Call"; *Merlin: The Complete First Season*, "A Remedy to Cure All Ills," BBC One, October 25, 2008, written by Julian Jones, directed by Ed Fraiman (qtd.).

26. *Merlin: The Complete First Season*, "The Beginning of the End," BBC One, November 8, 2008, written by Howard Overman, directed by Jeremy Webb; *The Adventures of Merlin: The Complete Second Season*, "The Witch's Quickening," BBC One, December 5, 2009, written by Jake Michie, directed by Alice Troughton; *The Adventures of Merlin: The Complete Second Season*, "The Fire of Idirsholas," BBC One, December 12, 2009, written by Julian Jones, directed by Jeremy Webb; *Third*, "The Crystal Cave."

27. *First*, "The Dragon's Call."

28. *Fifth*, "The Disir."

29. *Fifth*, "The Drawing of the Dark."

30. *First*, "The Beginning of the End."

31. *First*, "The Dragon's Call" (qtd.); *First*, "A Remedy to Cure All Ills"; *First*, "The Beginning of the End"; *Second*, "The Witch's Quickening"; *Second*, "The Fire of Idirsholas."

32. Aristotle, "De Generatione Animalium," in *Woman Defamed and Woman Defended: An Anthology of Medieval Texts*, ed. Alcuin Blamires (Oxford: Clarendon, 1992), 40; Alcuin Blamires, Introduction to *Woman Defamed and Woman Defended*, ed. Alcuin Blamires, 2; Eugene V. Gallagher, "A Religion without Converts? Becoming a Neo-Pagan," *Journal of the American Academy of Religion* 62 (1994): 853. While Aristotle referred to sexual roles, this was extrapolated into wider gender roles.

33. *First*, "The Dragon's Call"; *Merlin: The Complete First Season*, "The Moment of Truth," BBC One, November 22, 2008, written by Ben Vanstone, directed by David Moore; *First*, "Le Morte d'Arthur"; *Fourth*, "The Sword in the Stone, Part One."

34. *Fifth*, "The Kindness of Strangers."

35. This count excludes opponents who have no innate powers themselves but that use enchanted objects or beings.

36. *Merlin: The Complete First Season*, "Behind the Magic Part Two," DVD, produced by Mark Procter and Zoë Rushton (United Kingdom: Shine Limited and BBC Wales, 2010); *The Adventures of Merlin: The Complete Fourth Season*, Commentary for "The Secret Sharer" with Julian Jones and Richard Wilson, DVD (United Kingdom: Shine Limited and BBC Wales, 2013).

37. *The Adventures of Merlin: The Complete Third Season*, "The Tears of Uther Pendragon, Part Two," BBC One, September 18, 2010, written by Julian Jones, directed by Jeremy Webb.

38. *Fourth*, "The Secret Sharer"; *The Adventures of Merlin: The Complete Fourth Season*, "Lancelot du Lac," BBC One, November 26, 2011, written by Lucy Watkins, directed by Justin Molotnikov.

39. Larrington, 2.

40. Larrington 1, 5, 30, and 45. In particular, Geoffrey of Monmouth's *Vita Merlini* depicts her as Arthur's healer, and Chrétien de Troyes' *Erec* and *Yvain* mention her as simply Arthur's sister. Larrington notes that "from the early thirteenth century onwards, Morgan [sic] is always a comforting presence on the barge that bears Arthur away" (30), and that romances depicting Arthur's final battle usually depict Morgan "coming in person" to retrieve him (45).

41. *The Adventures of Merlin: The Complete Second Season*, "Secrets and Magic: A Behind the Scenes Look at Season Two: Episode 11," DVD, directed by Mark Procter, produced by Gillane Seaborne (United Kingdom: BBC Wales, 2011).

42. *Second*, "The Fire of Idirsholas"; *Third*, "The Crystal Cave"; *The Adventures of Merlin: The Complete Fifth Season*, "The Diamond of the Day, Part Two," BBC One, December 24, 2012, written by Julian Jones, directed by Justin Molotnikov.

43. *First*, "The Dragon's Call"; *The Adventures of Merlin: The Complete Third Season*, "The Coming of Arthur, Part Two," BBC One, December 4, 2010, written by Julian Jones, directed by Jeremy Webb.

44. *Merlin: The Complete First Season*, "The Gates of Avalon," BBC One, November 1, 2008, written by Ben Vanstone, directed by Jeremy Webb.

45. *First*, "The Gates of Avalon"; *The Adventures of Merlin: The Complete Second Season*, "The Nightmare Begins," BBC One, October 3, 2009, written by Ben Vanstone, directed by Jeremy Webb; *The Adventures of Merlin: The Complete Fourth Season*, "Aithusa," BBC One, October 22, 2011, written by Julian Jones, directed by Alex Pillai.

46. *First*, Commentary for "The Dragon's Call."

47. *First*, "The Beginning of the End"; *Second*, "The Witch's Quickening" (qtd.); *Fifth*, "The Disir" and "The Kindness of Strangers."

48. *The Adventures of Merlin: The Complete Fifth Season*, "Arthur's Bane, Part Two," BBC One, October 13, 2012, written by Julian Jones, directed by Justin Molotnikov.

49. *First*, "The Beginning of the End"; *Second*, "The Witch's Quickening."

50. Larrington, 1; Paxson, 118.

51. *Fifth*, "The Drawing of the Dark."

52. *Merlin: The Complete First Season*, "The Mark of Nimueh," BBC One, October 4, 2008, written by Julian Jones, directed by James Hawes; *Merlin: The Complete First Season*, "To Kill the King," BBC One, December 6, 2008, written by Jake Michie, directed by Stuart Orme; *Second*, "The Fire of Idirsholas"; *Third*, "The Tears of Uther Pendragon, Part Two"; *The Adventures of Merlin: The Complete Third Season*, "Queen of Hearts," BBC One, November 13, 2010, written by Howard Overman, directed by Ashley Way; *Fourth*, "The Darkest Hour, Part One"; *The Adventures of Merlin: The Complete Fifth Season*, "The Dark Tower," BBC One, November 10, 2012, written by Julian Jones, directed by Ashley Way; *Fifth*, "The Diamond of the Day, Part Two."

53. *First*, "Lancelot"; *First*, Commentary for "Lancelot"; *The Adventures of Merlin: The Complete Second Season*, "Lancelot and Guinevere," BBC One, October 10, 2009, written by Howard Overman, directed by David Moore; *Third*, "The Coming of Arthur, Part Two"; *Fourth*, "The Darkest Hour, Part One"; *The Adventures of Merlin: The Complete Fourth Season*, "The Darkest

Hour, Part Two," BBC One, October 8, 2011, written by Julian Jones, directed by Alice Troughton; *Fifth*, "The Dark Tower" (qtd.); *Fifth*, "With All My Heart."

54. Ulrike Bethlehem, *Guinevere, a Medieval Puzzle: Images of Arthur's Queen in the Medieval Literature of England and France* (Heidelberg: Winter, 2005), 320.

55. Bethlehem, 323.

56. *Ibid.*, 397 and 411 (qtd.).

57. *Fourth*, "Lancelot du Lac"; *Fifth*, "The Dark Tower" and "With All My Heart"; *The Adventures of Merlin: The Complete Fifth Season*, "The Hollow Queen," BBC One, November 24, 2012, written by Julian Jones, directed by Alice Troughton; *The Adventures of Merlin: The Complete Fifth Season*, "A Lesson in Vengeance," BBC One, November 17, 2012, written by Jake Michie, directed by Alice Troughton.

58. *Fifth*, "With All My Heart" (qtd.).

59. Bethlehem, 398 and 405 (qtd.).

60. *First*, "The Mark of Nimueh"; *First*, "To Kill the King"; *Second*, "Lancelot and Guinevere"; *The Adventures of Merlin: The Complete Second Season*, "The Sins of the Father," BBC One, November 14, 2009, written by Howard Overman, directed by Metin Huseyin.

61. *Third*, "Queen of Hearts."

62. *Fourth*, "The Darkest Hour, Part Two" (qtd.); *Fourth*, "The Sword in the Stone, Part Two."

63. This also suggests that racial equality is a non-issue in Camelot, as Guinevere is played by Angel Coulby, the only principal actor of color. The only acknowledgment of Guinevere's race is in a subtle joke created in naming her brother "Elyan," played by Nigerian-born British actor Adetomiwa Edun. In previous Arthurian literature, Sir Elyan bore the epithet "le Blanc."

64. *Fifth*, "The Diamond of the Day, Part One."

65. *Fifth*, "The Diamond of the Day, Part Two."

66. Bethlehem, 398 and 405.

67. Máirtín Mac an Ghaill, "The Irish in Britain: The Invisibility of Ethnicity and Anti-Irish Racism," *Journal of Ethnic and Migration Studies* 26 (2000): 139. The negative Celtic references are only the most obvious in relation to magic. The series also draws from non–European cultures to stigmatize its villains, as seen in the tendency for enemy soldiers to wear turbans and for sorceresses to have a combination of glamorous curls and dreadlocks.

68. Arthur Aughey, "Anxiety and Injustice: The Anatomy of Contemporary English Nationalism," *Nations and Nationalism* 16 (2010): 510.

69. Aughey, 507 and 510.

70. *Ibid.*, 510.

71. John Darwin, "Empire and Ethnicity," *Nations and Nationalism* 16 (2010): 392.

72. For a contemporary comparison, contrast the approval ratings of Queen Elizabeth II and Margaret Thatcher. Queen Elizabeth II, who has limited and little-used governmental power, had a 66 percent approval rating in 1998, rising to 90 percent in 2012 ("Ipsos MORI, Poll, Satisfaction with the Queen at record high," last modified June 15, 2012, http://www.ipsos-mori.com/researchpublications/researcharchive/2977/Satisfaction-with-the-Queen-at-record-high.aspx). Conversely, Margaret Thatcher's approval rating as Prime Minister "seldom rose above 40 per cent" (Sandra Wagner-Wright, "Common Denominators in Successful Female Statecraft: The Political Legacies of Queen Elizabeth I, Indira Gandhi, and Margaret Thatcher," *Forum on Public Policy* 2012.1 [2012]: 14).

73. *Fifth*, "The Hollow Queen."

Bibliography

The Adventures of Merlin: The Complete Second Season. Produced by Julian Murphy and Johnny Capps. United Kingdom: Shine Limited and BBC Wales, 2011, DVD.

The Adventures of Merlin: The Complete Third Season. Produced by Julian Murphy and Johnny Capps. United Kingdom: Shine Limited and BBC Wales, 2012. DVD.

The Adventures of Merlin: The Complete Fourth Season. Produced by Johnny Capps and Julian Murphy. United Kingdom: Shine Limited and BBC Wales, 2013. DVD.

The Adventures of Merlin: The Complete Fifth Season. Produced by Johnny Capps and Julian Murphy. United Kingdom: Shine Limited and BBC Wales, 2013. DVD.

Aristotle. "De Generatione Animalium." In *Woman Defamed and Woman Defended: An Anthology of Medieval Texts*, edited by Alcuin Blamires, 39–41.

Aughey, Arthur. "Anxiety and Injustice: The Anatomy of Contemporary English Nationalism." *Nations and Nationalism* 16 (2010): 506–524.

Bethlehem, Ulrike. *Guinevere, a Medieval Puzzle: Images of Arthur's Queen in the Medieval Literature of England and France*. Heidelberg: Winter, 2005.

Blamires, Alcuin, ed., with Karen Pratt and C.W. Marx. *Woman Defamed and Woman Defended: An Anthology of Medieval Texts*. Oxford: Clarendon, 1992.

Blamires, Alcuin. Introduction to *Woman Defamed and Woman Defended*, ed. Alcuin Blamires, 1–15.

Bradley, Marion Zimmer. *The Mists of Avalon*. New York: Knopf, 1982.

Brown, Callum G. *The Death of Christian Britain: Understanding Secularisation 1800–2000*. New York: Routledge, 2001.

Cumper, Peter, and Tom Lewis, eds. *Religion, Rights and Secular Society: European Perspectives*. Northampton, MA: Edward Elgar, 2012.

Darwin, John. "Empire and Ethnicity." *Nations and Nationalism* 16 (2010): 383–401.

Davie, Grace. "Religion in Europe in the 21st Century: The Factors to Take into Account." *European Journal of Sociology* 47 (2006): 271–296.

Davie, Grace. *Religion in Britain since 1945: Believing without Belonging*. Cambridge, MA: Blackwell, 1994.

Edge, Peter W. "Secularism and Establishment in the United Kingdom." In *Religion, Rights and Secular Society*, edited by Peter Cumper and Tom Lewis, 38–57.

Gallagher, Eugene V. "A Religion without Converts? Becoming a Neo-Pagan." *Journal of the American Academy of Religion* 62.3 (1994): 851–867.

Graves, Robert. *The White Goddess: A Historical Grammar of Poetic Myth*. New York: Farrar, Straus and Giroux, 1948. Amended and enlarged edition, 1966.

Ipsos MORI. "Ipsos MORI, Poll, Satisfaction with the Queen at record high." Last modified June 15, 2012. http://www.ipsos-mori.com/researchpublications/researcharchive/2977/Satisfaction-with-the-Queen-at-record-high.aspx.

Larrington, Carolyne. *King Arthur's Enchantresses: Morgan and Her Sisters in Arthurian Tradition*. New York: I.B. Tauris, 2006.

Merlin: The Complete First Season. Produced by Julian Murphy, Johnny Capps, Jake Michie, and Julian Jones. United Kingdom: Shine Limited and BBC Wales, 2010. DVD.

Mac an Ghaill, Máirtín. "The Irish in Britain: The Invisibility of Ethnicity and Anti-Irish Racism." *Journal of Ethnic and Migration Studies* 26 (2000): 137–147.

Paxson, Diana L. "Marion Zimmer Bradley and *The Mists of Avalon*." *Arthuriana* 9 (1999): 110–126.

Wagner-Wright, Sandra. "Common Denominators in Successful Female Statecraft: The Political Legacies of Queen Elizabeth I, Indira Gandhi, and Margaret Thatcher." *Forum on Public Policy* 2012.1 (2012). http://forumonpublicpolicy.com/vol2012.no1/archive/wagner.wright.pdf.

York, Michael. "Invented Culture/Invented Religion: The Fictional Origins of Contemporary Paganism." *Nova Religio: The Journal of Alternative and Emergent Religions* 3.1 (1999): 135–146.

York, Michael. "New Age Commodification and Appropriation of Spirituality." *Journal of Contemporary Religion* 16 (2001): 361–372.

Ne cherchez pas la femme

The Women of *Kaamelott*

TARA FOSTER

The recent spate of television shows with a medieval theme includes Alexandre Astier's cult series *Kaamelott*, a French retelling of Arthurian legends that ran from 2005 through 2009. The series, which takes place in fifth-century Britannia, presents a chiefly parodic vision of the court and its inhabitants for a target audience of adults.[1] Although English-language adaptations of the Arthuriad heavily outnumber French productions in all media, the popularity of *Kaamelott* proves once again the appeal of the myth and its central figure since, "over the fifteen centuries of his literary life [...] each age, each culture found in [King Arthur] an iconic figure embodying something significant for its society."[2] Of particular interest to this essay are the female figures who surround Arthur, for while the women of *Kaamelott* are seen rather less frequently than its men, they nonetheless play a crucial role in the life of the court. As Thelma Fenster has noted, "in spite of their extraordinary malleability from culture to culture and through the centuries, female Arthurian figures seem to arrive in each new work with a full set of already-givens that carry the freight of the problem that is woman."[3] Questions about the treatment and status of women surface frequently in the series, indicating the extent to which the Arthurian legend still serves as a powerful lens through which social constructs such as gender roles can be examined. In this essay, we will focus on the presentation of Guenièvre and other principal female figures in *Kaamelott* using Maureen Fries's categorization of Arthurian women as a framework, looking at ways in which the show subverts a number of the "already-givens" of various female characters and carves a space for female agency within its version of King Arthur's world.

Before turning to the representation of women in *Kaamelott*, let us first look briefly at the series as a whole since many readers might be unfamiliar with it. Viewers who tuned in to *Kaamelott* when it first aired in 2005 might have been surprised by the irreverence with which the legend is treated, particularly if they knew Eric Rohmer's *Perceval le Gallois* and Robert Bresson's *Lancelot du Lac*. In these earlier French Arthurian adaptations,

> the principal referent is not a pseudo-historical recollection of archaic military glory and political ascendancy, but some of the greatest works of medieval literature: above all, the 12th-century verse romances of Chrétien de Troyes, like the *Conte du Graal*, and the 13th-century prose romances of the Vulgate (or Lancelot-Grail) cycle, like the *Mort Artu*.[4]

Both of these films were quite serious in tone, and each one met with critical acclaim in large part for showcasing "the *auteur*'s understanding of his material."[5] *Kaamelott*, on the other hand, presents what at first appears to be the "lighter side" of Camelot and has no principal referent, drawing instead from a broad spectrum of sources in its reconstruction of the famed court. Like a number of modern adaptors of the legends, Astier sets his series in the Romano-British period, but he does not attempt to recreate the "historical" Arthur any more than he bases his adaptation on any single source, literary or cinematic. This is not to say, however, that Alexandre Astier is any less an *auteur* than Rohmer or Bresson, or indeed any other Arthurian author of any age: the scope of his inter-textual allusions attests to his familiarity with many permutations of the legends, and his deep involvement (both on and off screen) in the series marks *Kaamelott* as very much his personal creation.[6] In Astier's version of Arthur's world, most of the knights cannot even begin to live up to their glorious reputations, demonstrating instead a lack of competence, courage, and common sense that frequently enrages Arthur and renders his mission of finding the Holy Grail seemingly hopeless. For the spectators, however, the knights' many flaws are delightfully entertaining, particularly in the (decidedly comedic) first three seasons of the series. Arthur's exasperation with the majority of his knights and the growing schism between himself and Lancelot make his task increasingly burdensome, however, and the tone of the series grows progressively darker in the final three seasons.

As with his treatment of Arthur's entourage, Astier's portrait of the king himself departs significantly from textual tradition. Whereas the literary Arthur frequently occupies a secondary role while his knights pursue their adventures, the Arthur of *Kaamelott* remains center stage, and we often see him leading one or more of his knights on a quest. It is the king rather than Lancelot who inspires unwavering loyalty from Bohort, whose counterpart in *La Mort le roi Artu* steadfastly supports his cousin Lancelot in the war against Arthur. In the series, Bohort tries desperately to talk Lancelot out of destroying Britain by turning his back on the king, but the knight believes that he is better fit to rule the kingdom than Arthur: "'Vous n'êtes pas de taille à mener la Quête du Graal. Vous n'avez pas les épaules. De plus, [...] vous passez pour un faible et un laxiste auprès du peuple et des pays voisins'" ['You aren't up to leading the Quest for the Holy Grail. You don't have the shoulders for it. Furthermore, the people and neighboring lands take you for a weakling who is too lax'].[7] In the final episode of the sixth and final season of the series, Lancelot has taken over Kaamelott, burning the Round Table and rounding up Arthur's knights for execution, while the physically weakened Arthur flees to Rome.

As the preceding overview of the series suggests, *Kaamelott* focuses more on the king and his male entourage than it does on the female characters, a bias which the series has in common with many medieval romances. Roberta Krueger notes that most Old French verse romances "cast women more often as desired objects rather than as active subjects in chivalric adventures or quests."[8] Elsewhere she comments that with regard to many of the female figures in Chrétien de Troyes's *Chevalier de la Charrette*, for example, "their status and motivations are unexplained."[9] Regina Psaki finds that this description "also fits the female figures in the later compilation romances" produced in 13th-century Italy, since women in those texts "are still essentially corollary to the male hero, still extrinsic, and still more the bearers than the makers of meaning."[10] While *Kaamelott*'s women, like their literary forebears, are not the principal protagonists, their status and motivations do not always go unexplained, and the series does not limit women to the role of "bearers of meaning."

To help illuminate the presentation of female characters in *Kaamelott*, we will refer to the archetypes identified by Maureen Fries in her analysis of the various roles played by women in Arthurian literature. Like Krueger and Psaki, Fries finds that "Arthurian women are essentially ancillary to the male actors of that literary tradition," and she concludes that they "must therefore be considered in relation to the male heroic roles they complement or defy: as heroine, female hero or counter-hero."[11] In Fries's terms, a heroine performs a traditional, passive female role, serving as inspiration to the male hero and as confirmation of his valor; she also confirms the values of the patriarchal social order to which she is bound. She relies on the male hero to take action and does not demonstrate her own agency. A female hero, who might be a virgin like Lunete or a wife like Enide, is able to move and act of her own accord, albeit temporarily and in the service of male heroes and patriarchal values; her heroism is thus performed not in her own interests but in those of her male counterparts. Finally, a female counter-hero like Morgan le Fay is frequently sexually aggressive and dangerous, acting in her own interests and in ways that transgress patriarchal norms. Fries's paradigm serves as the foundation for the recent analysis of Arthurian women in Geoffrey of Monmouth by Fiona Tolhurst, who both underscores its continued value and proposes that it might be usefully modified. In particular, Tolhurst observes that "Donald L. Hoffman has already used the female figures in Sir Thomas Malory's *Le Morte Darthur* to demonstrate that Fries's categories can overlap" and that the paradigm "assumes that female figures never play traditionally male roles such as king or hero (in the traditional, male sense) and cannot act in their own interest without becoming dangerous or destructive to males," prompting us to allow for more fluidity within the paradigm.[12] All three of Fries's archetypes are present in *Kaamelott*, but as we shall see, characters in the series can also transcend categories, and one of Astier's most compelling characters shows traces of the female counter-hero that are presented as admirable rather than dangerous or threatening.

If we look at Guenièvre through Fries's framework, we might initially classify her as a heroine, a woman who performs "a traditionally identified, female sex-role [...] of preserving order (principally by forgoing adventure to stay at home)."[13] Her definition of herself as a creature whose primary purpose is to serve the needs of the king seems to exemplify her "submission to patriarchal custom."[14] As she says to Lancelot,

> **Guenièvre:** C'est juste que, des fois, j'aimerais pouvoir lui donner ce dont il a besoin.
> **Lancelot:** Je vous entends toujours parler de lui, de ses besoins.... Et vous?
> **Guenièvre:** Et moi?
> **Lancelot:** Oui, vous. De quoi est-ce que vous avez besoin?
> **Guenièvre:** Ben...qu'il aille bien.
> **Lancelot:** Non mais d'accord, mais vous...vous n'avez jamais envie de rien?
> **Guenièvre:** Si, qu'il aille bien.
> **Lancelot:** Bon, j'abandonne.
> **Guenièvre:** Qu'est-ce que vous voulez que je vous dise? Je suis Reine, je fais ce qu'on m'a appris : je m'occupe du bien-être du Roi [I.37, 236–7].
>
> [**Guenièvre:** It's just that, sometimes, I'd like to be able to give him what he needs.
> **Lancelot:** I always hear you talking about him, about his needs.... What about you?
> **Guenièvre:** Me?
> **Lancelot:** Yes, you. What do you need?
> **Guenièvre:** Well, for him to be well.
> **Lancelot:** Yeah, OK, but you...don't you ever desire anything?

Guenièvre: Yes, for him to be well.

Lancelot: Fine, I give up.

Guenièvre: What do you want me to say? I'm the queen, I do what I've been taught: I take care of the king's well-being.]

The queen's apparent lack of self-interest in her marriage underscores her role as heroine, "an object of exchange or an object of desire" who "valorizes the knights' honor with her approving presence as spectator."[15] In *Kaamelott*, tournaments do not take place and women are not present to observe any knightly adventures, but Guenièvre is on hand to observe and approve of Arthur's behavior with his many mistresses. When Lancelot points out that women other than the queen are tending to the king's well-being, she responds, "'Elles sont charmantes, je suis désolée de vous le dire! Et puis c'est flatteur d'avoir un époux qui a du goût pour les jolies femmes!'" ['They are charming, I'm sorry to tell you! And it's flattering to have a spouse who has a taste for pretty women!'] (I.37, 237). The series thus draws on both medieval literary tradition (the queen usually plays the role of adulteress) and contemporary French cultural stereotypes (men usually play the role of adulterer) in its comedic treatment of the couple. The satirical nature of the series invites viewers to ridicule Arthur's adultery and to spurn the queen's endorsement of it as absurd; such an invitation seems calculated to undermine rather than valorize Arthur and to indicate that what is often seen as "Guenevere's troublesome sexuality" in the textual tradition belongs to *Kaamelott*'s king instead.[16]

Like the characters of the knights and of the king himself, Arthur's relationship with his wife differs substantially in Astier's series from what we find in the medieval Arthuriad, and *Kaamelott*'s presentation of that relationship does much to rehabilitate the queen whose reputation is tarnished in many of her medieval (and modern) incarnations by her adulterous affair with Lancelot. In *La Mort le roi Artu*, for example, the affair is "identified as a principal cause of the Arthurian tragedy"[17]; in *La Tavola Ritonda*, "the medieval author (and, perhaps, audience) have no difficulty in identifying the flaw at the heart of the Arthurian dream of chivalry with Ginevra's transgressions."[18] In *Kaamelott*, by contrast, viewers would be hard pressed to point to Guenièvre and her actions as the source for the kingdom's destruction. Astier's parodic treatment of Arthur's story overturns a number of situations that viewers familiar with Arthur might expect to find, chief among them the love triangle between Lancelot, Guenièvre, and Arthur. Although adulterous desire does exist between the characters, it is much modified from its medieval configuration. It might be better described as a one-way chain rather than a triangle: Lancelot loves and desires Guenièvre, Guenièvre loves and desires Arthur, Arthur desires a bevy of official and unofficial mistresses and acquaintances and loves (and desires) his first wife,

Aconia, and his third wife, Mevanwi. Thus, the transgressive sexuality in the television series is demonstrated first and foremost by the male characters, and most particularly by Arthur himself, rather than by Guenièvre.

The series repeatedly demonstrates Arthur's aversion to having sex with Guenièvre, suggesting that the lack of a legitimate heir might not be the fault of the queen but of the king. We see Séli, Guenièvre's mother, give Arthur a fertility potion (brewed by Merlin and therefore useless, since he is as incompetent as the knights [I.23]); we hear Arthur tell a peasant girl who seems to be expecting his child that he doesn't have the heart to apply himself with Guenièvre (I.43); we watch Arthur invent excuses for why he cannot procure a Roman aphrodisiac for the queen (I.56); we observe her failed attempt to follow Demetra's advice on how to seduce Arthur (I.95); we listen to his increasingly implausible explanations for what he is doing in bed with four of his mistresses after claiming that he couldn't sleep with Guenièvre for fear of tiring himself out before a military campaign (III.44). Turning to Arthur's mistresses for advice, the queen confirms her suspicions that she won't get pregnant because the king has never done "what is necessary" to conceive a child with her:

> **Aelis:** C'est à vous de faire l'héritier!
> **Aziliz:** Mais allez-y, dites-nous ce que vous faites, on vous dira si c'est bon! [...]
> **Tumet:** Vous dites juste ce qui se passe en-dessous des hanches et au-dessus des genoux.
>
> *Guenièvre reste quelques secondes pensive.*
>
> **Guenièvre:** J'en était [*sic*] sûre que c'était par là que ça se passait! [...] Mais enfin, je ne comprends pas... Arthur a jamais essayé de faire ça avec moi!
> **Aelis:** Jamais, jamais? [...] La descendance est assurée [III.64, 314–316].

> [**Aelis:** It's your place to create the heir!
> **Aziliz:** But go ahead, tell us what you do, we'll tell you if it's good!
> **Tumet:** Just tell us what happens below the hips and above the knees.
>
> *Guenièvre remains pensive for a few seconds.*
>
> **Guenièvre:** I was sure that that's where it happened! But I don't understand... Arthur has never tried to do that with me!
> **Aelis:** Never ever? Descendants are guaranteed.]

Aelis and Aziliz go on to berate Arthur for his failure to ensure the bloodline and the political future of Logres, to which he can offer only a feeble defense: "'Pas un mot! Là, je risque mon titre! [...] Si, j'ai essayé mais...voilà. J'ai failli... j'ai failli dégobiller et puis c'est tout!'" ['Not a word! I'm risking my title here!

Yes, I did try but...there it is. I almost...I almost barfed and that's all there is to it!'] (III.64, 316). Arthur's sexuality rather than Guenièvre's, then, is transgressive, since he has failed to perform his conjugal duties for over ten years.

However, the pretext that Arthur offers to his mistresses proves to be disingenuous, for his reasons for not sleeping with the queen are much more profound. In the sixth season, *Kaamelott* returns us to Arthur's youth in Rome, where his mother Ygerne sent him to receive a military education. After a calculating Roman senator learns of Arthur's British origins and of the legendary sword that designates the king of the Britons, he decides to advance the young Arthur rapidly through the ranks in order to place him on the British throne as a Roman puppet. As a part of this scheme, the senator introduces Arthur to the lady Aconia, who is charged with educating the young soldier so that he might properly represent Rome in his new, illustrious station. The two fall in love, and although she is some twenty years his senior and married to Britain's current Dux Bellorum (a detail that she does not reveal to Arthur), they marry in secret. When Arthur tells her that a political marriage is being arranged for him with the daughter of the intransigent Léodagan of Carmélide, she grants Arthur permission to take a second wife, provided that he never consummate the marriage. Thus, we discover that Arthur's refusal to sleep with Guenièvre and the consequent lack of a legitimate heir to the throne stems from Arthur's loyalty to his first wife rather than his seeming revulsion for his current one. Although *Kaamelott* employs "the mechanisms by which our gaze is constantly directed back to the male protagonists" used in so many medieval retellings of the legend, mechanisms that "make it impossible to focus on the female characters independent of their male counterparts," the strictures imposed by Aconia late in the series redirect our gaze to the feminine.[19] Geraldine Heng studies the "disruptive gestures and energies, intrusions and interruptions that [...] ultimately point to a submerged second narrative interplaying with and often prompting the first, and marked by a recognisably feminine voice" in Malory's *Morte Darthur*,[20] and we might see Aconia's stipulations as an example of one such disruptive gesture in *Kaamelott*: despite its focus on the male characters, the undercurrent of women's involvement in shaping the series's narrative is, at least on occasion, shown to be far-reaching.

While Arthur's failure to sire an heir to the throne might be seen in a positive light as proof of his fidelity to Aconia, he later compounds his sexual transgression by having an affair with Mevanwi, the wife of the knight Karadoc. Arthur knows that his behavior with Mevanwi contravenes the will of the gods, for a seer whom he met during his time in Rome delivers a dire warning: "'Ne touche pas à la femme du Chevalier, Arthur. Les Dieux le prendront comme un affront'" ['Do not touch the knight's wife, Arthur. The gods will take it as an affront'] (III.87, 418). Furthermore, his behavior breaks one of Kaamelott's fundamental laws, a law of which he is well aware:

> **Mevanwi:** Donc, les Chevaliers ont le droit d'avoir un nombre de maîtresses incalculable sauf...
> **Arthur:** Sauf les femmes des autres Chevaliers. Sinon il y a combat à mort [III.93, 446].
>
> [**Mevanwi:** So, knights have the right to have an incalculable number of mistresses except...
> **Arthur:** Except the wives of other knights. Otherwise there's a fight to the death.]

When Mevanwi comes to his bed in the appropriately-titled episode "La Faute: Première Partie" [The Sin: Part One] (IV.8), Arthur reminds her of the dangers involved and then proceeds to commit the offense that will bring about an initial disintegration of the court.[21] His attempt to circumvent the law by officially exchanging wives with Karadoc cannot prevent the destruction that he has already set in motion. In one of her very rare appearances in the series, La Fée Morgane leads Arthur to a book of prophecies that indicates the gravity of Arthur's transgression, for it predicts a "siècle des larmes, hurlements, [...] panique, ruine, fin d'un monde. Sur Terre sans démons ni sorcières, vient Dieu des morts, solitaire des frayeurs, du Ciel à l'insulte la Réponse" [century of tears, screams, panic, ruin, end of a world. To an Earth without demons or witches comes the God of the Dead, lone figure of fright, from Heaven the Response to the insult] (IV.50). This same episode marks the first sighting of a shadowy, hooded figure who ultimately encourages Lancelot to undertake the destruction of everything that Arthur has built, calling Lancelot "l'élu, mais pas de ces dieux-là. [...] Vous êtes l'élu des seuls dieux qui remboursent le prix du sang versé. [...] Concentrez-vous, car votre tâche est considérable" [the chosen one, but not of those gods. You are the chosen one of the only gods that repay the price of spilled blood. Focus, for your task is considerable] (V.2). This figure, Méléagant, identifies himself as "La Réponse," reminding us of the prophecy and pointing the finger at Arthur rather than at Guenièvre for the impending devastation.

As for Guenièvre's role in the initial downfall of the kingdom, *Kaamelott* exonerates her almost entirely. Despite his retinue of official mistresses, Arthur inspires strong loyalty from the queen, whose devotion and naïveté prevent her from noticing Lancelot's interest in her, thus creating a stark contrast with the queen of the *Prose Lancelot*. In that medieval text, the queen clearly sees that Lancelot is in love with her but toys with him "por veoir coment ele [lo] porra metre a malaise, car ele cuide bien que il ne pansast d'amors s'a lui non" [to see how she could make him uneasy, for she truly believes that he has no thought of love for anyone other than herself], and when the knight dares not make the first move, the queen takes him by the chin and initiates their first kiss.[22] Likewise, Malory's Guinevere "is imperious, impulsive, and sometimes witty. She exercises her power" over Lancelot, which "labels her as capricious, cruel and arbitrary in the view of her husband and other knights."[23] Such a description does not apply to *Kaamelott*'s queen. In the first season, we witness Lancelot's first attempt to communicate his feelings for Guenièvre and her utterly oblivious response.

> **Lancelot:** Dans certaines contrées, un homme choisit une femme, l'épouse et ne connaît qu'elle jusqu'à ce que la mort les sépare.
>
> **Guenièvre:** Oui mais nous, ça n'a rien à voir, on est civilisés, on n'est pas des sauvages! Vous les voyez, les hommes d'ici, avec une seule femme toute leur vie? Ils auraient l'air de quoi?
>
> **Lancelot:** Moi je trouve ça bien. [...] J'aime une femme en secret et n'en aimerai pas d'autre jusqu'à ma mort.
>
> **Guenièvre:** Mais pourquoi en secret? Vous êtes pas plus moche qu'un autre! [...]
>
> **Lancelot:** Cette femme est l'épouse d'un personnage très important. Et qui plus est, d'un ami fidèle.
>
> **Guenièvre:** [...] Quitte à choisir qu'une seule femme, vous pouviez pas en prendre une libre? [I.37, 237–8].
>
> [**Lancelot:** In certain lands, a man chooses a woman, marries her and knows no one but her until death separates them.
>
> **Guenièvre:** Yes, but that has nothing to do with us, we're civilized, we're not savages! Can you see the men from here with just one woman their whole lives? How would that look?
>
> **Lancelot:** I think it's good. I love a woman in secret and will love none but her until my death.
>
> **Guenièvre:** But why in secret? You're no uglier than anyone else!
>
> **Lancelot:** This woman is the wife of a very important person. And what's more, of a loyal friend.
>
> **Guenièvre:** Able to choose just one woman, couldn't you take an available one?]

Guenièvre's total incomprehension of Lancelot's intent underscores her artlessness.[24] Whereas the queen of the *Prose Lancelot* easily interprets the knight's silence as an expression of love, the queen of the series repeatedly fails to decode Lancelot's words, even when he suggests that he can pay a discreet after-hours visit to her room during Arthur's absence and release her from her chastity belt so that she can have a pleasant night. Fundamentally loyal, she cannot even conceive of Lancelot's betrayal of Arthur.

The queen does finally come to know Lancelot's feelings for her and she does leave Arthur for him, but the series again portrays Arthur as the ultimate target for censure in adultery. At the end of the third season, Lancelot decides to abandon Arthur and the Round Table since he believes that the buffoons of the court are unfit to undertake the Quest for the Holy Grail: "'Je m'en vais et je vais me débrouiller tout seul. Et même tout nu et sur un pied, j'irai toujours cent fois plus loin que vous et votre risible compagnie de crétins'" ['I'm leaving and I'm going to get by on my own. And even totally naked and on one foot, I will still go a hundred times farther than you and your laughable company of cretins'] (III.99, 472). Before leaving, he charges Bohort to reveal to Guenièvre the secret of his love for her, a revelation that leaves her stunned. A mere two episodes later, the queen learns the secret of Arthur's feelings for Mevanwi, and this betrayal drives her to leave the court to join Lancelot. Just as her loyalty prevents her from seeing Lancelot's feelings for her, she is similarly blind to the signs of Arthur's betrayal. The penultimate clue finds her in the kitchens for a late-night snack when Mevanwi enters; she jokingly asks whether Mevanwi is there for an assignation, and Arthur's arrival just seconds later merely prompts her to comment on how peckish everyone seems to be. The next time Arthur is missing from her bed at night, however, she catches the pair kissing in the kitchens and cannot fail to understand: "'La femme d'un Chevalier.... La faute suprême.... Vous me noyez dans la honte....'" ['The wife of a knight.... The ultimate sin.... You're drowning me in shame....'] (III.100, 478). Thus, Arthur's actions push the queen into the arms of the man whose departure has destabilized the kingdom, but even after she joins Lancelot, she remains technically innocent of adultery while Arthur proceeds to consummate his relationship with Mevanwi. Guenièvre has no real opportunity to become an adulteress, much to her dismay (and to our amusement): Lancelot reveals that he has saved himself for her and hopes that she can show him what to do, and he refuses to ask one of his men for an explanation for fear of undermining his authority in their eyes. Still a virgin, Guenièvre expresses serious doubts about her relationship with Lancelot, who replies that, unlike Arthur, he will never let her leave: "Je préfère vous tuer de mes mains plutôt que de vous perdre" [I would rather kill you with my bare hands than lose you] (IV.54). The increasing madness of her new consort causes her to rejoice when Arthur rescues her from what everyone knows has devolved into her captivity, and she returns to the court without any recriminations or lingering suspicions about her involvement with the knight.

Astier's rehabilitation of Guenièvre gives her a greater moral authority than that of the king and his principal knight, and it allows her to transcend "the heroine's role as reflector of the male hero's values," values that the series has called into question.[25] We have seen that the sexual misconduct of Arthur and Lancelot lies close to the heart of what is rotten in this kingdom; as Cédric Briand remarks, "life in Kaamelott indeed favors male hyper-hetero-sexuality," and the conflicts that arise as a result help "pav[e] the way to dystopia."[26] Despite her seemingly passive acceptance of this "hypermasculinity," Guenièvre does at times criticize patriarchal customs that are disadvantageous to women, acting as a vehicle for questioning the gender status quo, both medieval and modern. A case in point is the question of polygamy. Given her comical tolerance of Arthur's traditional right to have countless mistresses, we are not surprised to hear the queen reject the ban on polygamy proposed by the woman-hating Répurgateur, an Inquisitorial figure who represents the worst kind of religious bigotry and hypocrisy: "'Moi, je dis que vous devriez vous débarrasser de ce type avant qu'il ne répande ses idées ridicules dans tout le pays'" ['I say you should get rid of that guy before he spreads his ridiculous ideas throughout the land'] (I.15, 104). In hindsight, we know that Arthur's refusal of the Répurgateur's proposal is fueled largely by his promise to Aconia not to sleep with Guenièvre, but our first viewing of the episode underscores what we initially interpret as Arthur's aversion to the queen and his reluctance to find himself with no other women in his life (and bed). The episode also makes us laugh when Père Blaise points out, "'C'est moderne'" ['It's modern'] (I.15, 103), for it turns Arthur's use of this phrase to justify his progressive policies against him. However, Guenièvre pushes Arthur's progressive tendencies even further by suggesting that women should be able to enjoy the same privileges as men: "'Par contre, il y a une chose à laquelle on pourrait réfléchir, c'est autoriser la polygamie pour les femmes'" ['On the other hand, there is something we could think about—allowing polygamy for women'] (I.15, 104). The king instinctively refuses, unable even to articulate his objection to the idea, and his laughable knee-jerk reaction again invites ridicule while it reminds us that the sexual double standard still exists today.

Guenièvre's position as a not entirely silent or submissive heroine evolves into something more like a female hero in Fries's terms as the series progresses, for she demonstrates her agency and has an impact on the community of Kaamelott. Later seasons emphasize the queen's moral qualities rather than her stupidity, although her stupidity is frequently used humorously to undercut Arthur's sexual morality and to denounce gender inequality at court. She again questions prevailing gender norms in the first season when she inquires about the absence of women from the Round Table: "'[I]l y a jamais de femmes à la Table Ronde? [...] Alors pour quelle raison est-ce que les femmes n'ont pas le droit d'être là?'" ['Are there never any women at the Round Table? Well, for what reason do women not have the right to be there?'] (I.59, 60). When she sees a figure who appears to be a woman leave the next meeting, Guenièvre applauds Arthur for taking her words to heart, unaware that the figure is in fact merely Calogrenant wearing the newly-invented kilt (with absurd origins). As one of Fries's heroines, we can expect the queen to be, like her medieval literary counterparts, "the instrument around whom the action turns, [...] carried off and imprisoned; fought for and defended; freed and returned home, and fought for again; all at the will of and/or agreement between the males in the tale."[27] Although she is imprisoned by Lancelot and freed by Arthur, her departure from the court is her own decision, and an abduction that takes place in the first season and in which she agrees to participate turns out to have been staged by her mother, Séli. Guenièvre thus proves her potential for action and mobility early on, and her behavior following her temporary split with Arthur moves her closer to the "wife-hero" whose "agency is 'selfless' in that it exists for patriarchal—male rather than female—purposes."[28] The effort of dealing with fools takes its toll on Arthur, as does the knowledge of all that Britain has cost him (the wife he loved, a child, the freedom to pursue his personal happiness), so he decides to give up his power and go looking for any illegitimate children that he might have fathered. In a gesture of loyalty and support, Guenièvre accompanies him on his new quest; joining her husband incidentally allows her to escape her overbearing father, whom she appointed regent upon Arthur's renunciation of the throne, but the primary purpose of her participation in the search is to help her husband. Like Enide in Chrétien de Troyes's romance, she recognizes danger to her husband when he and everyone else seem oblivious to it, but unlike Enide, Guenièvre herself comes to Arthur's rescue, striking down both of his attackers (V.6). Guenièvre's second act of heroism, her attempt to save Arthur's life when Lancelot comes to kill him, does not succeed, but using her body to deny the knight access to the king's bath constitutes another "assertion of real female prowess."[29]

Guenièvre seems to have inherited her capacity for action from her mother, Séli, one of the new faces to join *Kaamelott*'s familiar cast of Arthurian characters and one of the most compelling of the series's female figures. Although she most frequently plays the role of female hero in that her actions benefit the kingdom as a whole, she also goes beyond that role to act as a counter-hero furthering her own interests in defiance of the kingdom's norms. Unlike Fries's counter-heroes, however, Séli is portrayed as neither dangerous nor threatening since the norms that she undermines are presented as detrimental to the community. An invented character, Séli does not come with the "full set of already-givens" borne by characters like the queen and thus does not need to be redeemed before her agency manifests itself.[30] Séli is far more assertive than her daughter, and she plays a more active role in the political life of the kingdom, both at Arthur's court and in her husband's federated territory of Carmélide. Léodagan counts Séli herself as one of the forces behind their daughter's departure from the court to live with Lancelot: "Vous lui avez bourré le mou à cette gosse, voilà tout! C'est les femmes qui décident, les femmes c'est des guerrières, les femmes si, les femmes mi...." [You filled that kid's head with nonsense, that's all! Women are the ones who decide, women are warriors, women this, women that] (IV.1). Séli refuses to remain silent and thus refuses the role of heroine, claiming instead the right to voice her opinions as loudly as does her husband. When Séli's father-in-law criticizes the outspokenness that she has also passed on to Guenièvre, the series immediately undercuts his criticism. He remarks, "'Et voilà...la fille et la femme qui interdisent au père.... La décadence est en marche'" ['And so...the daughter and the wife who forbid the father.... Decadence is under way'] (I.63, 87), but his admission that people are fleeing Carmélide for Kaamelott proves that his people reject his outdated, chauvinistic attitudes.

Séli's "fruitful use of tongue and wit" often serves the interests of Arthur's kingdom, as in the many cases when she tries to facilitate the production of an heir, but it also serves her own purposes.[31] For example, whereas the literary sources show various male figures abducting the queen, Séli herself orchestrates the "abduction" of Guenièvre in "a gesture of feminine resistance to an ideology that circulates women as objects."[32] Séli enacts her plan for personal financial gain because she never benefitted from her own marriage to Léodagan, a marriage that was itself the result of an abduction from which he, and he alone, made a tidy sum. When the men hear that a band of Picts has taken the queen, Léodagan confides to Arthur that they are dealing with formidable adversaries: when he abducted Séli (a Pictish princess) for a ransom of ten thousand gold pieces, they paid him twenty thousand to keep her. It seems likely that the Picts were anxious to be rid of an opinionated princess who did not know her "place." Unable to speak Pictish and unaware of the subterfuge, Arthur and Léodagan must rely on Séli to negotiate the ransom and accept her suggestion that they begin with an offer of thirty thousand gold pieces, which she then shares with Guenièvre since "'le pognon, faut que ça circule'" ['cash needs to circulate'] (II.13, 96). Séli's insistence on the circulation of money rather than women, her critique of a sexual economy that treats women as disenfranchised objects of exchange and barter, underscores that she "holds values which are not necessarily those of the male culture in which she must exist," making of her a female counter-hero in Fries's terms.[33] Such disenfranchisement is the target of the series's satire on more than one occasion, however, rendering Séli's counter-heroism laudable rather than wrongful.[34]

Séli's involvement in governance means that she is frequently at Léodagan's side when questions of policy are discussed, and she is generally presented as (at least) his equal; she thus also transcends the category of female hero with regard to the duration of her agency. While Arthurian romance "allows [the wife-hero] only intermittent access to those activities ordinarily (and 'properly') associated with males," Séli enjoys permanent access to such activities.[35] She demonstrates her awareness of her importance in managing the affairs of her kingdom, as well as a distinct lack of deference for her husband: "Tout le monde le sait, en Carmélide, que vous êtes une patate. Seulement, ça passe parce que les gens savent que c'est moi qui tire les ficelles dans l'arrière boutique" [Everyone in Carmélide knows that you're a dolt. However, it's accepted because people know that I'm pulling the strings behind the scenes] (IV.2). Moreover, her political role is not limited to joint decisions with Léodagan. She displays her aptitude for political intrigue on a number of occasions, such as when she concocts a plan for putting a Guenièvre who has been impregnated by Lancelot back on the throne, a scheme that is doomed to failure by Lancelot's sexual naïveté; on other occasions, she acts as the official representative of the Pictish people, fulfilling a "rol[e] ordinarily attributable only to men."[36] Ygerne, Arthur's mother, similarly represents the people of Tintagel, and we see the two women discussing the merits of allowing their two peoples to remain at war, since it occupies the men and stops them from degenerating into "'des gros cons'" ['total jerkoffs'] (II.39, 245). Despite her relatively minor role, Ygerne is also a female hero, and her decision to send Arthur to be educated in Rome constitutes an example of "the feminine origin of many of the enabling conditions for activity."[37] After all, Arthur points out to a Guenièvre who is shocked to learn of her husband's upbringing, only his time in Rome allows him to take power in Logres and to oust all but one garrison of Roman troops. Given the military superiority of the Empire, "'Pourquoi est-ce qu'ils acceptent de laisser un Roi "local" fédérer les Clans bretons? [...] Parce que je suis de chez eux. Si j'avais pas fait mes classes à Rome, il n'y aurait pas de Kaamelott. Pensez-y'" ['Why do they accept letting a "local" king federate the British clans? Because I'm from there. If I hadn't studied in Rome, there would be no Kaamelott. Think about it'] (II.62, 84–5). Ygerne's initiative thus proves to be of paramount importance, for her agency in political decision-making results in wide-ranging benefits for all of the federated clans.

On the supernatural plane, the Dame du Lac plays a vital role as a female hero who enables Arthur's success, whereas the Fée Morgane, the female counter-hero par excellence in Fries's framework, occupies a surprisingly marginal position. La Dame du Lac, who represents a council of Celtic goddesses, establishes the founding principles of Arthur's Kaamelott: the fellowship of the Round Table and the Quest for the Holy Grail. Wace tells us in his *Roman de Brut* that the construction of the Round Table was Arthur's idea, but in *Kaamelott*, Bohort tells us that "'Sire Arthur a eu une révélation de la Dame du Lac qui lui a ordonné la construction d'une table légendaire autour de laquelle il devrait réunir les

Chevaliers de Bretagne pour organiser la Quête du Graal'" ['Lord Arthur had a revelation from the Lady of the Lake who ordered him to have built a legendary table around which he must unite the knights of Britain in order to organize the Quest for the Holy Grail'] (I.3, 35).[38] Thus, the Dame provides both the material object most closely associated with the kingdom as well as its raison d'être; the directive to undertake the Quest does not come directly from the (masculine) One God but via the (feminine) goddesses and their (feminine) agent. In addition to his overarching mission, the Dame also directs Arthur in a plethora of smaller adventures by which he proves his prowess and maintains his reputation. As Sue Ellen Holbrook finds for the Dame's counterpart in Malory, she "receives her light from the male heroes whose orbit she moves in," so when Arthur abandons the quest for the Holy Grail, the Dame's light effectively goes out[39]; she is expelled from the council of goddesses and becomes mortal, barely able to function in the terrestrial realm. Although her potential for female heroics is much diminished after her fall from grace, the Dame du Lac is still able to perform one last deed whose positive impact on the court is immeasurable, for it is she who reminds Lancelot of the healing spell that he uses to save Arthur's life after the king has slit his own wrists. Her healing spell counters the evil magic that holds Lancelot enthralled, but her enemy is the otherwordly Méléagant rather than her sometime literary opponent, the Fée Morgane. As for the traditional counter-hero Morgane, the series subverts viewers' expectations of her malevolence: she appears on only two occasions, and her villainy is non-existent.

While the series features a number of female heroes and a praiseworthy counter-hero in the character of Séli, two figures qualify as the dangerous counter-hero identified by Fries. In her study of contemporary Arthurian fiction written by female authors, Roberta Davidson finds that Arthur is often at the center of a romantic triangle and that he is

> fought over by his wife Guenevere and a lover who is most often Morgan le Fay, Nimue, Morgause, or some sexual and/or sorceresstic equivalent. The equation works with predictable regularity: when Guenevere is good, Morgan (et al.) is bad. When Guenevere is bad, Morgan (et al.) is good.[40]

Whichever character is "bad" is depicted as "over-sexed, selfish, [and] scheming," because "[s]omeone, apparently, has to bear the [...] sexual blame" for the transgression at the root of the fall of the Arthurian world.[41] Since both Guenièvre and Morgane are good in *Kaamelott*, the sexual blame falls (partially) on another woman: Mevanwi. She displays some of the "over-sexed" qualities of the villainous women in Arthurian fictional texts, coming to Arthur's bed to seduce him as soon as she learns that Guenièvre has left to join Lancelot.

> **Arthur:** Attendez.... Euh, ça fait des semaines qu'on se voit dans les coins de porte. Il faut pas qu'on se touche parce que soi-disant ça va se voir. Et là, patatras,

vous dégringolez directe dans la piaule et il faudrait tout boucler en un quart d'heure?

Mevanwi: Ça vous plait pas?

Arthur: C'est pas que ça me plait pas, c'est que je voudrais comprendre, c'est tout.

Mevanwi: J'en ai marre d'attendre. Je… je… je pense à des choses.

Arthur: Des choses? Quelles choses?

Mevanwi: Ben, ben je vous montrerai sur pièce, on va pas faire un coloc' là!

Arthur: Écoutez! Vous faites la maligne là, genre… euh… j'arrive, je provoque. […]

Mevanwi: Je sais mais j'ai envie [IV.8].

[**Arthur:** Wait…. Um, we've been seeing each other in doorways for weeks. We can't touch each other because supposedly it will be seen. And now, whammo, you hop straight into bed and we have to seal the deal within a quarter of an hour?

Mevanwi: Don't you like it?

Arthur: It's not that I don't like it, it's that I'd like to understand, that's all.

Mevanwi: I'm sick of waiting. I… I… I'm thinking about things.

Arthur: Things? What things?

Mevanwi: Well, I'll show you on the spot, let's not have a confab!

Arthur: Listen! You're playing the little devil now, like… uh… here I am, I'm being provocative.

Mevanwi: I know, but I want it.]

She suggests sexual role-play scenarios, and although Arthur calls her out on her behavior, he does not refrain from having sex with her, again underscoring that he has not been duped into sleeping with her but has consciously chosen to do so despite being warned by the gods not to touch her. Because Mevanwi schemes to get revenge on Arthur after he annuls the wife swap and reclaims Guenièvre, we can classify her as a dangerous counter-hero according to Fries's categorization. Like some of her medieval forebears, she learns the arts of sorcery in order to exact retribution in the form of a deadly potion. When that rather transparent plot fails, she seduces the jurisconsult, the legal expert who has been summoned to adjudicate the restoration of Guenièvre as Arthur's wife; she thus acquires and then destroys by magic the annulment of her marriage to Arthur, thereby making herself the queen once more, with her husband Karadoc as the official regent. She also tries to convince her husband to decree that the queen will wield power rather than her consort, but her strategy backfires, leaving her with less influence than before. The "negative themes of hypersexuality, misused power and ugliness masquerading as beauty" all apply to Mevanwi, for while she is beautiful, Perceval comments frequently on her unattractiveness[42]; idiotic but pure of heart, perhaps Perceval sees more than the viewers can. Even given these negative attributes of the female counter-hero, the series does not totally vilify Mevanwi, who later comforts and supports Guenièvre, and the most over-sexed character in the show is Arthur himself.

Mevanwi is not the only threatening female counter-hero in the series; Anna, Arthur's sister, also occupies this position, corresponding to later medieval versions of Morgan le Fay that cast the nefarious sorceress as Arthur's sister, although she makes very few appearances in the series and thus far plays a very minor role in Astier's story. She resembles the literary Morgan in her ruthlessness and sexuality much more strongly than Mevanwi does, for "[e]ven in marriage, she exhibits an indifference to patriarchal values and a sexual freedom unknown and unknowable to the female hero or the monogamous heroine."[43] Her husband affirms that "la moitié de la Bretagne lui est passée dessus" [half of Britain has taken a turn on top of her] (IV.69), and she appears to Arthur in a vision, asking if he would enjoy sleeping with her and assuring him, "Vous y viendrez" [You'll come around] (VI.9), thus opening up the possibility of the incestuous conception of Mordred in the continuation of the story. Anna despises Arthur, blaming him for the death of her father, the duc de Gorlais. Like Mevanwi, she tries to kill the king, but her plot is much more subtle and is thwarted only by the heroics of Guenièvre. Dangerous and iron-willed, Anna nonetheless does not play the role of "the most extreme villain" of *Kaamelott*.[44] That title more properly belongs to Lancelot, who also resolves to murder Arthur and ends the series by setting the Round Table on fire and willfully slaughtering his former brothers-in-arms; to Méléagant, who convinces Lancelot that it is his duty to carry out such merciless bloodshed; and perhaps to Arthur himself, whose poor decisions result in the arrival of Méléagant and unleash his destructive energies.

In her recent study of Arthurian fiction for young adults, Fiona Tolhurst adds two categories of female protagonist to Fries's framework: those of "tragic heroine" (a character who "loses the love of her life because of a flaw in her character and/or an error in judgment") and "female warrior-hero" (a character who "wields a sword or other weapon for her own ends").[45] Although we do not see any tragic heroines in *Kaamelott*—Mevanwi loses Arthur because of his fears about the prophecy rather than her own shortcomings—we do get a brief glimpse of a character with the potential to become a female warrior-hero. In one of the pilot episodes, a woman in full armor joins a meeting of the Round Table with a recommendation from Léodagan, but she realizes that Arthur assumed that the recommendation was for her to be taken on as a maid. Arthur replies, "Non, pas forcément. [...] Dans ma tête, il vous recommandait pour que je vous marie à un chevalier, par exemple" [No, not necessarily. In my head, he was recommending you so that I would marry you to a knight, for example], a position that confirms her assessment of how the knights view women: "C'est ce que je dis. Bonniche" [That's what I said. A maid] ("Chevalier femme" [Woman knight]). When Arthur accepts her presence since nothing in Père Blaise's archives says that it's forbidden, the knights grumble and complain about the damage this will do to their reputation. Edern points out that her gender should not be the sole basis for her exclusion from the company: "Attendez, attendez, le problème, c'est que je suis une fille, c'est ça? C'est pas juste, ça! J'ai jamais porté de rose, hein? J'aime pas les fleurs, je me lave pas, et je pisse dans mon armure comme vous!" [Wait, wait, the problem is that I'm a girl, is that right? That's not fair! I've never worn pink! I don't like flowers, I don't bathe, and I piss in my armor like you!]. When she leads the charge to throw out the misogynist, homophobic Archbishop of Bologna who questions whether the Vatican will continue to support the knights in the Quest for the Holy Grail if they are nothing but a bunch of "tapettes" [fairies], she earns their respect and an invitation to consider them her brothers. As the meeting comes to a close, however, she loses the respect of all of the knights except Bohort when she opens her bag and offers them some cakes that she made, a change of heart that again mocks the men and the chivalric ethos for its insistence on a sharp division of gender roles that denies women the opportunity to contribute to society in a broader capacity.

Whether Edern would have played the role of female warrior-hero remains a mystery, for the character never appears in the series proper, nor, with the exception of Guenièvre's impromptu defense of Arthur, do we see other women take up arms and fight in *Kaamelott*. We might attribute this absence in part to the series's French origins, for the two most recent English-language Arthurian series, the BBC's *Merlin* (2008–2012) and Starz' *Camelot* (2011), feature a number of female characters who are proficient in the use of weapons and prepared to participate in battle. Virginia Blanton remarks that audiences "are more attuned to females who are capable of fighting alongside the men (like Trinity in *The Matrix*)" and "the American desire for strong female characters has been repeatedly met by lead characters in television and cinema, such as Xena, Warrior Princess; Buffy the Vampire Slayer; and The Bride in *Kill Bill*."[46] *Merlin*'s Gwen, Morgana, Morgause, and Isolde and *Camelot*'s Guinevere are thus part of a significant Anglo-American corpus of female action heroes that also includes characters such as *Robin Hood*'s Marian and Djaq and *Torchwood*'s Gwen back to *The Avengers*' Diana Peel, among many others. In contrast, French cinema and television are characterized more by "la rareté des femmes d'action" (the scarcity of female action heroes)[47]; that the category of female warrior-hero is not needed to characterize *Kaamelott*'s women tends to confirm this observation. However, as Jennifer Edwards establishes, although the women of *Merlin* and

Camelot can (and sometimes do) fight alongside the men, they are more likely to remain out of harm's way, and the shows "hold up for admiration and emulation examples of women who are passive, or emphasize capable women at their moments of helplessness."[48] The frustration felt by viewers of *Merlin* and *Camelot* when "the programs pay lip service to feminist themes without allowing for feminine success"[49] and that felt by viewers of *Kaamelott* who long to see a more prominent role for women demonstrate the continued difficulty of knowing "how to handle a woman" in Arthurian adaptations.[50] Indeed, the longevity of Fries's basic framework points to the pervasiveness of perceived notions about the roles women can or should play, but the need to expand or allow for more fluidity in its categories suggests a positive shift in those perceptions.

As we have seen, while women still tend to occupy an adjunct position in *Kaamelott*, they nonetheless exhibit agency that has a significant impact in Arthur's kingdom. The series may not allow them much room to tell their own stories and may not present them in entirely non-traditional roles, but it does share some of the features of contemporary feminist Arthurian fiction, such as the "creation of complex female characters who defy stereotypes" and the "portrayal of female characters as protagonists of stories that are not solely romantic ones."[51] The activities of the Dame du Lac, Ygerne, and Séli, for example, do not involve the pursuit of love, and none of the principal female characters are one-dimensional. As Andrew Elliott explains, "the episodic format of both the [medieval] romance and the TV series affords the ability to flesh out a character by the depiction of their everyday lives rather than solely by their epic actions,"[52] and despite the series's focus on its male characters, the women of Kaamelott are fleshed out enough for us to see Guenièvre develop from a heroine to a female hero, to see the counter-hero Mevanwi redeemed by her support of Guenièvre, and to see the counter-hero Séli consistently presented as admirable rather than dangerous. Furthermore, Arthur's policies as a "roi progressiste" [progressive king] include the rejection of a number of customs that abuse and demean women, including the refusal to rape the women of conquered peoples (I.87, III.14), the refusal to punish adulterous wives (II.19), and the refusal to have dancing girls on display at gatherings of local chieftains (I.16). Venec, the black marketeer, tries to convince Arthur that the dancers will facilitate his efforts at federating the clans represented by the chieftains: "'Je vous garantis qu'avec soixante paires de miches qui gigotent en même temps, vous fédérez ce que vous voulez derrière!'" ['I guarantee you that with sixty pairs of buns jiggling at the same time, you can federate whatever you want!'] (I.16, 115). The image of the female dancer who performs for and arouses a group of warriors strongly evokes the scene of Igraine dancing in John Boorman's film *Excalibur*, a "powerful male-friendly vision" of Arthur's world,[53] but Astier's Arthur sets his court apart from that of Boorman's narrative: "'Chez nous, les femmes sont pas traitées comme ça. [...] J'aime pas l'image que ça donne de nous'" ['Women aren't treated like that here. I don't like the image that gives of us'] (I.16, 115). The king recognizes that the objectification of women remains common currency among many of his countrymen but firmly rejects that attitude, and Astier invites his audience to do the same. Thus, in addition to creating a space for female agency, the show demonstrates a respect for women in general and denigrates those characters (the Répurgateur, Léodagan's father) who fail to show it. When Ygerne comes to visit her son, she praises his stance on women: "'Vous avez mis fin à ces coutumes misogynes et c'est une chose qui doit être portée à votre crédit'" ['You have put an end to those misogynist customs and you must be given credit for that'] (II.27, 175). As viewers of *Kaamelott*, it is tempting to say something similar to Alexandre Astier and to hope that future Arthurian adaptors might aim to surpass him.

Notes

1. The series changed both in format and in tone as it developed over six seasons, with earlier seasons featuring much shorter and more comedic episodes than later seasons.
2. W. R. J. Barron, "*Bruttene Deorling*: an Arthur for Every Age," in *The Fortunes of King Arthur*, ed. Norris J. Lacy, Arthurian Studies LXIV (Cambridge: D. S. Brewer, 2005), 47–65 (47).
3. Thelma S. Fenster, "Introduction," in *Arthurian Women*, ed. Thelma S. Fenster (New York: Routledge, 2000), xvii-lxiv (xx).
4. Jeff Rider, Richard Hull, and Christopher Smith, with Michael Carnes, Sasha Foppiano, and Annie Hesslein, "The Arthurian Legend in French Cinema: Robert Bresson's *Lancelot du Lac* and Eric Rohmer's *Perceval le Gallois*," in *Cinema Arthuriana: Twenty Essays*, ed. Kevin J. Harty, rev. ed. (Jefferson, NC: McFarland, 2002), 149–162 (150).
5. Rider et al., "The Arthurian Legend in French Cinema," 151.
6. *Kaamelott*'s creator and writer, Astier's other credits in the series include director, producer, editor, composer, and principal actor (he stars as Arthur). In addition to the short film and ten pilot episodes that precede the six seasons of the series, Astier has produced eight companion graphic novels and plans to bring even more installments of the story to fruition. Announced continuations of the story include a trilogy of feature films and a series of short stories entitled *Kaamelott: Résistance* to bridge the gap between the television series and the cinematic offerings. *Kaamelott: Résistance* might also be televised (http://tvmag.lefigaro.fr/programme-tv/article/people/76409/alexandre-astier-kaamelott-va-revenir.html). For a discussion of the ways in which Astier plays with his sources and claims authority for his new creation, see Tara Foster, "*Kaamelott*: A New French Arthurian Tradition," *Arthurian Literature* XXXI (2014): 185–201. For a discussion of how Astier forges links between the society he has created and that of his audience, see Tara Foster, "*Kaamelott*'s Global Fifth Century," forthcoming in *Arthuriana* 25.1 (March 2015).
7. "La Dispute," *Kaamelott*, season III, episode 99, dir. Alexandre Astier (Montreal: Alliance Vivafilm, 2011), DVD; Alexandre Astier, *Kaamelott: Livre III, texte intégral* (Paris: Éditions SW Télémaque, 2010), 472. Subsequent references to the third season of the series will refer to this text, and references to the first two seasons will cite the dialogue from Astier's other published scripts: *Kaamelott: Livre I, première partie* (Paris: J'ai lu, 2012); *Kaamelott: Livre I, deuxième partie* (Paris: J'ai lu, 2012); *Kaamelott: Livre II, première partie* (Paris: J'ai lu, 2012); *Kaamelott: Livre II, deuxième partie* (Paris: J'ai lu, 2012). Dialogue from the first three seasons will thus be cited by season, episode and page number. Readers should note that the dialogue produced by the actors differs on occasion from the published dialogue. References to episodes from seasons four through six, for which published scripts are not yet available, will be cited by season and episode number only. Translations from the French are my own.

8. Roberta L. Krueger, "Questions of gender in Old French courtly romance," in *The Cambridge Companion to Medieval Romance*, ed. Roberta L. Krueger (Cambridge: Cambridge University Press, 2000), 132–49 (137).

9. Roberta L. Krueger, "Desire, Meaning, and the Female Reader: The Problem in Chrétien's *Charrete*," in *Lancelot and Guinevere: A Casebook*, ed. Lori J. Walters, Arthurian Characters and Themes 4 (New York: Garland, 1996), 229–45 (233).

10. Regina L. Psaki, "'Le donne antiche e'cavalieri': Women in the Italian Arthurian Tradition," in Fenster, *Arthurian Women*, 115–31 (117).

11. Maureen Fries, "Female Heroes, Heroines and Counter-Heroes: Images of Women in Arthurian Tradition," in *Popular Arthurian Traditions*, ed. Sally K. Slocum (Bowling Green, OH: Bowling Green University Press, 1992), 5–17.

12. Fiona Tolhurst, *Geoffrey of Monmouth and the Feminist Origins of the Arthurian Legend*, Arthurian and Courtly Cultures (New York: Palgrave Macmillan, 2012), 7–8.

13. Fries, "Female Heroes," 6.

14. *Ibid.*, 9.

15. Krueger, "Desire, Meaning, and the Female Reader," 232.

16. *Ibid.*, 240.

17. Norris J. Lacy, "The Evolution and Legacy of French Prose Romance," in Krueger, *The Cambridge Companion to Medieval Romance*, 167–82 (172).

18. Psaki, "'Le donne antiche,'" 119.

19. *Ibid.*, 117.

20. Geraldine Heng, "Enchanted Ground: The Feminine Subtext in Malory," in Fenster, *Arthurian Women*, 97–113 (97).

21. As noted above, *Kaamelott* concludes with the destruction of the Round Table and its knights at the hands of Lancelot; we see Arthur, still weak after his attempted suicide, escape from Logres and the devastation carried out by Lancelot as a closing title announces, "Bientôt, Arthur sera de nouveau un héros" [Soon, Arthur will again be a hero] (VI.9). Although we cannot yet know the full scope of Astier's reinvention of the legendary king, it would appear that Fortune's Wheel has already taken Arthur full circle and is on the rise for a second time.

22. *Lancelot du Lac*, ed. Elspeth Kennedy, Lettres Gothiques, 2d ed. (Paris: Le Livre de Poche, 1991), 890. My translation.

23. Elizabeth Edwards, "The Place of Women in the *Morte Darthur*," in *A Companion to Malory*, eds. Elizabeth Archibald and A. S. G. Edwards, Arthurian Studies XXXVII (Cambridge: D. S. Brewer, 1996), 37–54 (50).

24. The depiction of Guenièvre on the cover of the published script (*Kaamelott: Livre I, deuxième partie*) wearing a white gown and holding a lamb also seems calculated to emphasize her innocence.

25. Fries, "Female Heroes," 7.

26. Cédric Briand, "*Kaamelott*'s Paradox: Lancelot between Subjugation and Individuation," forthcoming in *Arthuriana* 25.1 (March 2015).

27. Fries, "Female Heroes," 7–8.

28. *Ibid.*, 12.

29. *Ibid.*, 10.

30. Fenster, "Introduction," xx. Although the ghost of Guinevere's mother does make an appearance in the fifteenth-century *The Awntyrs off Arthur*, the character of Séli is essentially Astier's creation.

31. Fries, "Female Heroes," 12.

32. Krueger, "Desire, Meaning, and the Female Reader," 237.

33. Fries, "Female Heroes," 12.

34. In an earlier episode, we witness another pointed mockery of the sexual appropriation of women without their consent. Lancelot summons one of Arthur's subjects and announces to her, "'J'ai le plaisir de vous annoncer que le Roi Arthur vous a designée comme nouvelle maîtresse, faisant de vous une des femmes les plus importantes du Pays'" ['I have the pleasure of informing you that King Arthur has designated you as his new mistress, making you one of the most important women in the country'] (I.48, 303). When she refuses the designation, Lancelot is outraged: "'Nombre de demoiselles à votre place vendraient leur mère pour se voir offrir un tel honneur! [...] On

ne vous demande pas! C'est un ordre!'" ['Many young women in your place would sell their mother to be offered such an honor! You're not being asked! It's an order!'] (I.48, 304). Arthur does not impose his desire on the unwilling woman, however, and it is not until she has heard his motivations for choosing her, questioned him about the terms of her "appointment," and received satisfactory answers that she agrees to the arrangement; she therefore becomes an informed and willing participant rather than a sexual object with no voice or will.

35. Fries, "Female Heroes," 10.

36. *Ibid.*, 12.

37. Heng, "Enchanted Ground," 97.

38. Judith Weiss, *Wace's "Roman de Brut/ A History of the British": Text and Translation*, rev. ed. (Exeter: Short Run Press, 2002), vv. 9747–52, 244–5.

39. Sue Ellen Holbrook, "Nymue, Chief Lady of the Lake, in Malory's Le Morte Darthur," in Fenster, *Arthurian Women*, 171–190 (187).

40. Roberta Davidson, "When King Arthur is PG 13," *Arthuriana* 22.3 (2012): 5–20 (13).

41. Davidson, "When King Arthur is PG 13," 13.

42. Fries, "Female Heroes," 13.

43. *Ibid.*, 12.

44. *Ibid.*, 14.

45. Fiona Tolhurst, "Teaching Girls to Be Heroic?: Some Recent Arthurian Fiction for Young Adults," *Arthuriana* 22.3 (2012): 69–90 (70).

46. Virginia Blanton, "'Don't worry, I won't let them rape you': Guinevere's Agency in Jerry Bruckheimer's 'King Arthur,'" *Arthuriana* 15.3 (2005): 91–111 (92–3).

47. Raphaëlle Moine, *Les femmes d'action au cinéma* (Paris: Armand Colin, 2010), 8.

48. Jennifer C. Edwards, "Casting, Plotting, and Enchanting: Arthurian Women in Starz's *Camelot* and the BBC's *Merlin*," forthcoming in *Arthuriana* 25.1 (March 2015).

49. Edwards, "Casting, Plotting, and Enchanting."

50. Maureen Fries, "How to Handle a Woman, or Morgan at the Movies," in *King Arthur on Film: New Essays on Arthurian Cinema*, ed. Kevin J. Harty (Jefferson, N.C.: McFarland, 1999), 67–80.

51. Ann F. Howey, *Rewriting the Women of Camelot: Arthurian Popular Fiction and Feminism*, Contributions to the Study of Science Fiction and Fantasy 93 (Westport, CT: Greenwood Press, 2001), 113.

52. Andrew B. R. Elliott, "The Charm of the (Re)making: Problems of Arthurian Television Serialization," *Arthuriana* 21.4 (2011): 53–67 (64).

53. Fries, "How to Handle a Woman," 79.

Bibliography

Astier, Alexandre. *Kaamelott: Livre I, première partie.* Paris: J'ai lu, 2012.

_____. *Kaamelott: Livre I, deuxième partie.* Paris: J'ai lu, 2012.

_____. *Kaamelott: Livre II, première partie.* Paris: J'ai lu, 2012.

_____. *Kaamelott: Livre II, deuxième partie.* Paris: J'ai lu, 2012.

_____. *Kaamelott: Livre III, texte intégral.* Paris: Éditions SW Télémaque, 2010.

Barron, W. R. J. "*Bruttene Deorling*: an Arthur for Every Age." In *The Fortunes of King Arthur*, edited by Norris J. Lacy, 47–65. Arthurian Studies LXIV. Cambridge: D. S. Brewer, 2005.

Briand, Cédric. "*Kaamelott*'s Paradox: Lancelot between Subjugation and Individuation." *Arthuriana* 25.1 (forthcoming in March 2015).

Blanton, Virginia. "'Don't worry, I won't let them rape you': Guinevere's Agency in Jerry Bruckheimer's 'King Arthur.'" *Arthuriana* 15.3 (2005): 91–111.

Davidson, Roberta. "When King Arthur is PG 13." *Arthuriana* 22.3 (2012): 5–20.

Edwards, Elizabeth. "The Place of Women in the *Morte Darthur*." In *A Companion to Malory*, edited by Elizabeth Archibald and A. S. G. Edwards, 37–54. Arthurian Studies XXXVII. Cambridge: D. S. Brewer, 1996.

Edwards, Jennifer C. "Casting, Plotting, and Enchanting: Arthurian Women in Starz's *Camelot* and the BBC's *Merlin*." *Arthuriana* 25.1 (forthcoming in March 2015).

Elliott, Andrew B. R. "The Charm of the (Re)making: Problems of Arthurian Television Serialization." *Arthuriana* 21.4 (2011): 53–67.

Fenster, Thelma S. "Introduction." In *Arthurian Women*, edited by Thelma S. Fenster, xvii-lxiv. New York: Routledge, 2000.

Foster, Tara. "*Kaamelott*: A New French Arthurian Tradition." *Arthurian Literature* XXXI (2014): 182–201.

_____. "*Kaamelott*'s Global Fifth Century." *Arthuriana* 25.1 (forthcoming in March 2015).

Fries, Maureen. "Female Heroes, Heroines and Counter-Heroes: Images of Women in Arthurian Tradition." In *Popular Arthurian Traditions*, edited by Sally K. Slocum, 5–17. Bowling Green, OH: Bowling Green University Press, 1992.

_____. "How to Handle a Woman, or Morgan at the Movies." In *King Arthur on Film: New Essays on Arthurian Cinema*, edited by Kevin J. Harty, 67–80. Jefferson, N.C.: McFarland, 1999.

Heng, Geraldine. "Enchanted Ground: The Feminine Subtext in Malory." In *Arthurian Women*, edited by Thelma S. Fenster, 97–113. New York: Routledge, 2000.

Holbrook, Sue Ellen. "Nymue, Chief Lady of the Lake, in Malory's Le Morte Darthur." In *Arthurian Women*, edited by Thelma S. Fenster, 171–90. New York: Routledge, 2000.

Howey, Ann F. *Rewriting the Women of Camelot: Arthurian Popular Fiction and Feminism*. Contributions to the Study of Science Fiction and Fantasy 93. Westport, CT: Greenwood Press, 2001.

Kaamelott, Livres I-VI. DVD. Directed by Alexandre Astier. Montreal: Alliance Vivafilm, 2011.

Krueger, Roberta L. "Desire, Meaning, and the Female Reader: The Problem in Chrétien's *Charrete*." In *Lancelot and Guinevere: A Casebook*, edited by Lori J. Walters, 229–45. Arthurian Characters and Themes 4. New York: Garland, 1996.

_____. "Questions of gender in Old French courtly romance." In *The Cambridge Companion to Medieval Romance*, edited by Roberta L. Krueger, 132–49. Cambridge: Cambridge University Press, 2000.

Lacy, Norris J. "The Evolution and Legacy of French Prose Romance." In *The Cambridge Companion to Medieval Romance*, edited by Roberta L. Krueger, 167–82. Cambridge: Cambridge University Press, 2000.

Lancelot du Lac. Edited by Elspeth Kennedy. Lettres Gothiques, 2d ed. Paris: Le Livre de Poche, 1991.

Moine, Raphaëlle. *Les femmes d'action au cinéma*. Paris: Armand Colin, 2010.

Psaki, Regina L. "'Le donne antiche e'cavalieri': Women in the Italian Arthurian Tradition." In *Arthurian Women*, edited by Thelma S. Fenster, 115–31. New York: Routledge, 2000.

Rider, Jeff, Richard Hull, and Christopher Smith, with Michael Carnes, Sasha Foppiano, and Annie Hesslein. "The Arthurian Legend in French Cinema: Robert Bresson's *Lancelot du Lac* and Eric Rohmer's *Perceval le Gallois*." In *Cinema Arthuriana: Twenty Essays*, edited by Kevin J. Harty, 149–62. Rev. ed. Jefferson, NC: McFarland, 2002.

Tolhurst, Fiona. *Geoffrey of Monmouth and the Feminist Origins of the Arthurian Legend*. Arthurian and Courtly Cultures. New York: Palgrave Macmillan, 2012.

_____. "Teaching Girls to Be Heroic? Some Recent Arthurian Fiction for Young Adults." *Arthuriana* 22.3 (2012): 69–90.

Weiss, Judith. *Wace's "Roman de Brut/ A History of the British": Text and Translation*. Rev. ed. Exeter: Short Run Press, 2002.

Homosexuality in Television Medievalism

Torben R. Gebhardt

Homosexuality has been depicted in U.S. television series since the 1970s. However, programs tended to limit the role of gay men to mere effeminate drag queens. ABC's *Marcus Welby, M.D.* even degraded homosexuality to a mental illness (Season 4 Episode 22, original air date Feb. 20, 1973) and sparked immediate protest by gay groups and mental health professionals alike.[1] In the course of the decades that followed, homosexuals were given more and more attention in mass media. Nevertheless, it took until 1997 to find a homosexual main character in a television series. ABC's *Ellen* had its eponymous protagonist coming out in "The Puppy Episode" (Season 4, episodes 22 and 23, OAD April 30, 1997) which was an enormous ratings success and resulted in an Emmy win for the episode's script, co-written by leading actress Ellen DeGeneres. Regardless of this success the show was cancelled after another season of decreasing ratings and increasing focus on gay issues. This more serious tone was for many viewers the main reason why the show ceased to be funny.[2]

Nevertheless, *Ellen* meant a huge step forward for gay and lesbian visibility in America's mass media and was arguably the reason why another series with a gay, and this time male, main character was greenlit. NBC's *Will and Grace* was a huge success and ran for eight seasons. Centering on Will Truman, a successful lawyer, his best friend Grace Adler and their respective friends Jack and Karen, the show featured two male gay characters in leading roles on the small screen for the first time in history. With Will being depicted as responsible and level-headed, while Jack was impulsive and extremely emotional, the show's main characters embodied a spectrum of homosexual expression. While gay issues played an important role on the show, they remained non-threatening to heteronormativity, excluding Will's sexuality or using it almost solely as comic relief. Jack's depiction emphasizes this superficial treatment of the subject, for he is presented "not so much likeable as laughable."[3] The only relationship that is given any attention is that of Will with his heterosexual best friend Grace, which made Will—as iconic "gay best friend" —more palatable for a wider audience. The same is true for Ellen whose social environment is predominately heterosexual.[4]

In the case of gay male identities it is noteworthy that patriarchal notions of masculinity are rarely challenged when a homosexual is introduced as an ongoing character in a series. Rather, gay masculinity often equals what is understood as the desirable heterosexual male with sexuality elided or occluded. Fred Fejes put his finger on it when he said that "to be a gay male in today's world one would be young, white, Caucasian, preferably with a well muscled, smooth body, handsome face, good education, professional job, and a high income."[5] Thirteen years later this observation still rings true as can be seen in shows such as *Modern Family*, *Glee* or *The New Normal*. Additionally, U.S.-television's representation of homosexuality often reinforces traditional values such as family and monogamy.[6] *The New Normal* is a prime example of this. The series revolves around a long-term homosexual couple, David and Bryan, whose wish for a baby is fulfilled through the help of a surrogate mother. In their relationship the roles are clear cut, with David representing a more male stereotype, whose interests lie for instance in sports, while Bryan is the classic effeminate gay, dedicating a large part of his time to his appearance and loving to dress up for Halloween. These strategies of representation make homosexuality non-threatening to an audience that holds heteronormative values.

In 1999, the series *Queer as Folk*, produced by the British public service television Channel 4, revolutionized homosexuality on television with its unapologetic depiction of today's gay scene. *Queer as Folk* was unique in that it did not shy away from showing gay men and women engaging in sexual activities, from mere kissing all the way to penetrative sex. Although reactions differed and part of the Press, sponsors and viewers reacted with criticism, as did some queer critics who were afraid of the overly sexual message the show sent about the gay scene, *Queer as Folk* became a roaring success.[7] The next year the series was adopted for the North American market by the American network Showtime and the Canadian network Showcase and ran for five seasons. While the North American series still offered quite explicit views of gay sexuality, they were more concerned with making *Queer as Folk* appealing to a wider audience by staying within heterosexual gender norms. For example, an effeminate lesbian character gives birth on the show, while her more masculine (yet not what would be considered "butch" in the gay scene) partner cheats on her.[8] The aforementioned gay stereotype of "young, white, Caucasian, preferably with [...] a high income"[9] was also the norm on the show, keeping it from alienating the main focus group, namely young, white and well-off viewers.[10] Jane Arthurs is correct when she argues that the television landscape readily accommodates marginal groups "albeit by their ability to pay."[11] While *Queer as Folk* stayed gender-normative, it intentionally transgressed sexual borders by turning the cameras on gay characters between the sheets. This made the show more daring than other dramas and turned it into what can be termed "provocative entertainment,"[12] for a focus group. Such transgressions shock while kindling curiosity. While its notion of gender-identities remain, for the most part, normative, *Queer as Folk* crossed a border by showing sexually active homosexuals on an otherwise straight television landscape.

The same can be said about *The L Word*, also produced by Showtime from 2004 to 2009, which represents a kind of lesbian counterpart to *Queer as Folk*. This series applied the same conservative gender norms and was likewise successful. Additionally, the main cast consisted almost entirely of attractive young white women (often referred to as "femmes"), making the show more attractive for a male heterosexual market and excluding a more masculine type of homosexual woman (often referred to as "butch"). Both *Queer as Folk* and *The L Word* were consequently criticized for showing characters that are not representative of the broad spectrum of homosexual identity and expression.[13] Nevertheless, the success of *Queer as Folk* and *The L Word* represented a decisive step forward for gay visibility in television.

However, homosexuality has not yet played a noticeable part in the increasing number of television shows with a historical theme. That is not to say that there have not been a number of rather explicit scenes between characters of the same sex. *Rome* and *Spartacus* had their fair share of, foremost lesbian, sex scenes. *Spartacus* even went as far as introducing two gay gladiators in a relationship. In the sector of medieval television, homosexual sex-scenes may be featured, but same-sex relationships have been largely avoided. In their annual report on "Where We Are On TV" the GLAAD (Gay & Lesbian Alliance Against Defamation) listed two shows with medieval topics incorporating recurring homosexual characters: *The Borgias*, which is set at the far end of the Middle Ages (late 15th century), and *Game of Thrones*.[14]

This paper explores how gay characters are represented in medieval television by focusing on one of the most successful series on American television at the moment, *Game of Thrones* (HBO 2011–2014; production is ongoing). As a neomedieval series, *Game of Thrones* shows a medieval simulacrum with present concerns as its focus. The show uses straight characters' attitudes toward same-sex relationships to mark them as positive or negative characters and, similarly to gay characters in non-medieval television series, the two main gay characters Renly Baratheon and Loras Tyrell, re-inscribe heteronormative values of masculinity and gender roles. This paper uses *Game of Thrones* to draw conclusions about neomedievalism as well as about the importance of "provocative entertainment" to modern perceptions and representations of homosexuality.

"Authenticity" and Neomedievalism

George R. R. Martin categorized his novels as much as the television show based on them as medieval fantasy.[15] Comparable to Tolkien's *Lord of the Rings* or the hugely popular MMORPG[16] *World of Warcraft*, it is set in a completely imagined world inhabited by fantastical elements such as giants, dragons and magic that invokes the medieval.[17] The world of *Game of Thrones* may be different from that of the historical Middle Ages, but the period has been the most important source of information for George R. R. Martin. As he himself said in a recent interview, he is "drawing largely on medieval England, medieval Scotland, [*sic*] some extent medieval France."[18] The construction of Martin's world, however, is mainly achieved through the help of iconic images of the Middle Ages. These are combined with mythical elements taken from medieval literature to create a potpourri that represents not the historical Middle Ages but a new version of the medieval, a neomedieval world.

The term "neomedievalism" has been the subject of an extensive debate lately, largely focused on problems of clear definition. Scholars characterize neomedievalisms—such as fantasy novels and videogames—as simulations of the Middle Ages, usually made with modern technology, that conform more to audience-expectations than any notion of "historical authenticity."[19] However, elements such as the inclusion of fantastical elements as for instance dragons in such media actually *aligns* them with "authentic" medieval literature, which itself incorporates tales of fabulous beasts, including dragons.[20] We find dragons, for instance, as much in such famous literary works as *Beowulf*, the *Nibelungenlied* and *Le Morte d'Arthur* as we do in historical accounts like the *Anglo-Saxon Chronicle* which speaks of "fiery dragons (fyrenne dracan) ... flying in the air" for the year 793.[21] What today is called a "mythical creature" was without doubt a "wondrous beast" in the Middle Ages, yet it was part of the medieval world. *Game of Thrones* may eschew fidelity to historical *events* of the Middle Ages, but it certainly contains features "authentic" to medieval genres. Calling *Game of Thrones* and other such works of neomedievalism "inauthentic" ignores their use of a wider medieval repertoire. The mythic and romantic genres of the Middle Ages are key shapers of contemporary neomedievalisms, which are themselves modern simulacra of a medieval zeitgeist. Dragons are as much an "authentic" and iconic medieval symbol as knights and kings.

The audio-visual nature of television reflects the technological focus of neomedievalism. It is therefore the series *Game of Thrones*, more so than the novels, that creates a simulation of Martin's medievalism, showing every physical detail in HD on the small screen. The creation of a prevailing medieval mood is accomplished by applying images that encapsulate the viewer in the iconic Middle Ages without binding him/her to a historically predetermined sequence of events. Paradoxically, although the overall imagined nature of the work categorizes it as an inauthentic (i.e., "not real") rendering of the past, it attempts in many aspects to be true to its inspiration. Tom Holland, in an article for *The Guardian*, pointed out several aspects of *Game of Thrones* that are inspired by the Middle Ages and drew parallels between historical characters and their counterparts on the show, coming to the conclusion "that there are sequences where the invented world of Westeros can seem more realistic than the evocations of the past to be found in many a historical novel."[22] Once more we should ask the question of what kind of authenticity, and what kind of medievalism, the creators were aspiring to. For instance, the easiness with which characters in *Game of Thrones* sever limbs of even armored opponents with their swords is rather unlikely to reflect medieval practice correctly; yet it is reminiscent of literary works such as *Le Morte d'Arthur*.[23] The iconic medieval battle with all its blood and gore is therefore taken directly from the literary Middle Ages. Accordingly, ostensibly inauthentic features of *Game of Thrones* can be termed just as authentic as the connections that Tom Holland drew between historical persons and characters on the show. Regarding neomedievalism as indifferent toward historical authenticity ignores its indebtedness to medieval imagination. With that caveat in place, I retain the term "neomedievalism" as useful to describe *Game of Thrones.*

Medieval Homosexuality

Homosexuality in the historical Middle Ages is not easy to study. First of all, the medieval period is no coherent timespan and in many ways the early Middle Ages are closer to Antiquity than to the later Middle Ages. Moreover George R.R. Martin's epic is not constrained to a specific time within the medieval period. The large scale of his capital city as well as the focus on tournaments and elaborate heraldry point to the later Middle Ages. On the other hand, more remote settings such as Winterfell, and the way justice is dispensed point to the early or high Middle Ages. The image of the medieval potpourri comes to mind and is reminiscent of other neomedieval works such as Tolkien's *Lord of the Rings* or *World of Warcraft.*

A paucity of sources on homosexuality in the medieval era, compounded with the fact that clear categories of "heterosexual" or "homosexual" did not exist, makes modern analyses that place "gay" opposite "straight" anachronistic.[24] It needs to be stressed that same-sex intercourse was, nevertheless, seen as unnatural and unlawful in the medieval era. The designation "homosexual" was simply not used for a person engaging in a same-sex physical relationship. The prevailing perception was that the person had committed a crime, both against society and nature.[25]

A look at penitentials, guidelines for confessors regarding sins and their penance, from the Early Middle Ages reveals that same-sex intercourse demanded penance, but was in some cases, like in the penitential of Pope Gregory III, even less sinful than a priest who went hunting.[26] The Dominican monk Thomas Aquinas wrote in the thirteenth century that homosexuality was one of the unnatural vices, like bestiality and masturbation.[27] The condemnation is directed toward the act and not toward a sexual identity that deviates from the norm. The role a person took during the act could even be gender-determining. Penance was then ordered with disregard for the person's biological sex. *The Penitential of Theodore*, written in the later seventh century, for instance, states: "Sodomites shall do penance for seven years, and the effeminate man as an adulteress."[28] The effeminate man is stripped of his sex and gendered as a woman due to his role during intercourse. Something similar can be seen in the case of John Rykener, who was a transvestite prostitute in the late fourteenth century and often went by the name of Eleanor Rykener. The court has obvious difficulties in labeling the culprit with only one gender: "Rykener further confessed that [he] went to Beaconsfield and there, as a man, had sex with a certain Joan, daughter of John Matthew, and there also two foreign Franciscans had sex with him as a woman."[29]

Rykener is described according to his deed, either as a woman or a man. Here, just as in the example before, gender roles are clear cut and may even override the participant's sex. Whoever takes the receiving part during intercourse does not only act as a woman but actually becomes one in the view of the prosecutors. Sexual-identity categories such as transvestism or bisexuality seem not to have been available at the time. Still, to say that the Middle Ages had absolutely no concept of sexual identity is probably untrue and is opposed by many scholars.[30] W. Mark Ormrod proposed in his study on Edward II's sexuality the concept of "degeneracy" as an appropriate medieval view on homosexuality. Same-sex intercourse was regarded as "against kind" and carried with it other inadequacies, most strikingly that men were denied their masculinity and subsequently feminized. Furthermore, since the sin was not only regarded as a lack of restraint but a moral deficiency, it was often associated with other offences against the norm: leprosy, tyranny and heresy to name only a few.[31]

It is important to stress that intercourse was the focus of condemnation and not affection between members of the same sex, which appears to have been widely accepted in the Middle Ages and often more so than it is today. Richard I Lionheart, while at the French royal court, shared a bowl and a bed with King Philipp II without being condemned by contemporary chroniclers as "degenerate." This relationship was diplomatic and courtly rather than romantic.[32] However, hostility toward and intolerance for same-sex affection/intimacy or for same-sex intercourse increased from the twelfth century onward, condemning physical and mental homosexuality alike. This intolerance soon found its way into theological and legal writings.[33]

Homosexuality in Game of Thrones

Neomedieval works do more than simply create a simulacrum of the past by means of modern technology for today's audience: "the *neomedieval* simulacrum self-consciously interests itself in present concerns."[34] *Game of Thrones* interests itself in the way contemporary—not medieval—society deals with corporate culture, religious clashes, and, as explored here, homosexuality. Specifically, the television series belies homosexuality's contemporary, televised function as provocative entertainment. It becomes a neomedieval simulacrum, portrayed within the restrictions of a medieval world, but not as medieval homosexuality. Forbidden and intriguing, this neomedieval homosexuality is offered to viewers through a keyhole glance at isolated, transgressive occurrences.

Intercourse is prominently featured in television's *Game of Thrones*. Female-female sexual encounters are shown (rather than lesbian homosexuality) and these are cast as performances for male, heterosexual pleasure. In season one episode seven we see the character of Ros, a prostitute, having sex with another female prostitute while the brothel's owner Peter Baelish watches and later instructs them on how to be more pleasing for their future customers. He even assigns the roles of man and woman to them (invoking medieval perspectives) and thereby establishes that the intercourse is not about lesbian desire, but rather a training session for the women to improve their skills for their next male customer. In episode two of the first season a similar situation of same-sex erotic contact for instructive purposes occurs. Doreah, a servant, instructs the newlywed Queen Daenerys on how to satisfy her husband. Both scenes should be seen in the same light as what has been established above about sex on *Queer as Folk*. It is "provocative entertainment," an isolated occurrence that is intriguing without pushing the borders too far (by positing intimate love between two women) and therefore still reassuring the viewer in his or her heteronormative values. The only instance where lesbian homosexuality is approached in a more general sense is in season three episode seven when Margaery Tyrell tells Sansa Stark that women like all different kinds of men and some even other women. While the remainder of the conversation strongly implies that Margaery is drawing from personal experience, homosexuality is of little importance to the scene. It only serves to establish Margaery as the much more experienced and intelligent protagonist, not as a lesbian. Homosexuality as a subject is only present in its male version in *Game of Thrones*. Foremost is the relationship between two influential—though not main— characters in the kingdom: Renly Baratheon and Loras Tyrell.[35]

The televised adaptation retains most of the original features of these two gay characters. Young, white, Caucasian, good-looking and wealthy, Loras Tyrell and Renly Baratheon fit perfectly the stereotype that Fred Fejes has described. Renly's "Rainbow Guard" strikes obvious comparisons to the symbol of the LGBT (lesbian, gay, bisexual and transgender) community, the "Rainbow Flag." Yet, the show makes considerable alterations to the novels' treatment of homosexuality that are worth exploring in detail. The most striking is the explicitness with which the series depicts the two characters' homosexual relationship: the novels only hint at the sexual orientation of Loras and Renly. For instance, when Jaime Lannister encounters Loras Tyrell in the capital city, Loras draws his sword to attack Jaime's companion whom he believes to be responsible for Renly's death, to which Jaime replies: "Now sheathe your bloody sword, or I'll take it from you and shove it up some place even Renly never found."[36] Such a direct implication of a homosexual relationship between Renly and Loras could, in the context of the novels, be based on false rumors, and Loras' vengeful reaction could simply demonstrate a close friendship with Renly, though there are other remarks and hints that make it quite apparent that George R. R. Martin intended these characters to be in a homosexual relationship. The books never explicitly describe sexual acts between the two: the chapters in each novel are written from the point of view of various characters, and neither Renly nor Loras ever function as one of these POV characters. Any sexual encounters between the two would be hidden from POV characters and therefore from the audience. However, the television series brings male-male sex acts on-screen.

Loras and Renly's relationship, or at least rumors of it, is surprisingly well-known throughout Westeros. In episode four of the second season two random soldiers of house Lannister talk about Loras Tyrell. One calls him "pretty" and doubts his qualities as a swordsman, because "[h]e's been stabbing Renly Baratheon for years and Renly ain't dead." The other guard does not seem to need further explanation to understand the implication, suggesting that the relationship between Renly and Loras may not be openly acknowledged, but as a rumor is known even to the lower classes.

One way that the writers of the *Game of Thrones* series use homosexuality is as a means to define heterosexual characters. While homosexual acts are not accepted as equal to heterosexuality, "positive" characters are open-minded about homosexuality while "negative" characters condemn it along the lines of the medieval denigrations described above. The neomedieval setting, with its prevailing hostility against homosexuality, serves as a background that allows the creators to cultivate contemporary liberal perspectives of homosexuality. A scene that has no counterpart in the books in episode six of the third season features a conversation between Olenna Tyrell, head of House Tyrell, and Tywin Lannister, head of House Lannister. When Tywin mentions rumors of Loras Tyrrell's sexual preference Olenna says her grandson is a "sword swallower through and through." Yet, she sees it as a very natural thing that young men "have a go at each other" and congratulates Tywin on his restraint after he denies having ever been involved with another man. Tywin Lannister on the other hand calls the act "unnatural" and an "affliction," a view that corresponds with Ormond's theory of the medieval conception of homosexuality as "degeneracy." Tywin's daughter Cersei, queen until the king dies in the middle of season one, similarly calls Renly a "known degenerate." Her son, who becomes king after his father's death, shares his mother's view and even considers making Renly's "perversion," as he calls it, punishable by death. While Olenna's comments don't mark her as fully embracing a homosexual identity for her grandson—she regards his sexual preference as nothing more than a phase—others approach it with more hostility, considering it a permanent illness. Interestingly, the Lannisters use Renly Baratheon's homosexual behavior to establish him as unfit for the role of king, mirroring medieval attitudes, while Olenna represents "positive," contemporary "acceptance" of the acts as natural. Rather than marking same-sex intercourse as a deficiency, such scenes have the effect of establishing the accusing characters as hostile, narrow-minded, "backwards" and "medieval."

Other remarks, by Peter Baelish or Jaime Lannister for instance, do not condemn homosexuality in the same way. They do not refer to the physical or psychological degeneration of homosexuality but attempt to take advantage of the secrecy surrounding it. In episode five of season one, Renly and Peter bet on the outcome of a joust between Loras and Gregor Clegane, an exceptionally strong knight. While Renly advises Peter to buy some friends with the money should he win, Baelish asks Renly "And when will you have your friend?" The clear implication serves its purpose of unsettling Renly Baratheon, who immediately sits down and does not say another word. Although their relationship appears to be known and occasionally incites contempt, overall, neither Renly Baratheon nor Loras Tyrell seem to be treated disrespectfully by other members of the nobility. Eddard Stark, one of the most respected characters in *Game of Thrones*, greets Renly Baratheon in episode three of the first season with genuine fondness, either unaware of his sexual orientation or indifferent toward it. The latter seems to be a fitting description for the general attitude toward same-sex love in George R. R. Martin's world. Its existence appears to be widely known, yet it is only rarely a subject of discussion. On the few occasions when it is of importance it either marks characters as hostile and unpleasant or it is used as a rhetorical weapon to unsettle known homosexuals.

The importance of religion in shaping attitudes toward homosexual acts in *Game of Thrones* differs significantly from medieval society. In its structure and hierarchy, the dominant religion is quite similar to Christianity. The center of this faith is located in the kingdom's capital King's Landing with the High Septon as a quasi–Pope at its head. Although the so-called "Faith of the Seven" is prominent in this society there is no scene linking it to homosexuality. Neither is any stated opinion on Renly and Loras' relationship religiously motivated. While one would have expected an involvement in the matter by the Church of Westeros, this is in concordance with the overall tone of the series and displays once more the show's neomedieval interest in present concerns over representations of a medieval worldview or cultural moment. While every character is described as being devout, at least in public, the predominant religion is of almost no importance to the storyline. Instead, other religions appear and increase in relevance over the course of the series, but seem to have no position on homosexuality either. Only Tywin Lannister's remark on homosexuality as something "unnatural" bears some resemblance to the point of view of the medieval Church. However, it seems to be only his personal opinion rather than the dominant mode in Westeros.

Masculinity and Identity

In *Game of Thrones'* neomedieval society, traditional gender roles are for the most part intact. Men fight and rule while women breed children and take care of the household. The only exceptions are Brienne of Tarth and Arya Stark, both of whom take up arms in combat, as well as Cersei Lannister who rules as Queen Regent and Daenerys Targaryen who becomes Queen in the lands east of Westeros. However, the first two are repeatedly marked as exceptions to the rule, just like Daenerys who claimed direct rule only after her husband died, while Cersei does a rather poor job. Therefore, none of these characters challenge established gender roles, especially since Daenerys is repeatedly referred to as "mother" of her people, thereby stressing her gender. One might think that a homoerotic, homosexual relationship could work to undermine a clear division of society by gender roles; however, in this series, the characterizations of Loras Tyrell and Renly Baratheon rather entrench heteronormative values.

As Peter Baelish turned his female prostitutes into a "man" and "woman" when they practiced, and as medieval law turned men into women based on sexual role, so too does the series perforate Loras and Renly along such lines. However, the aspect of "medievalism" lends a bit more nuance to these roles.

In the series, Loras incorporates masculine features and is depicted as the driving force in the relationship, representing a strong and able prototype of a knight. Renly on the other hand is continuously feminized through the subordinate nature of his role in the relationship. Although Loras and Renly are both effeminate in appearance, it is only Loras who is an excellent swordsman and jouster, whereas Renly is never seen wielding a sword or a lance. He is even ridiculed by his brother and king Robert for all the "balls and masquerades" (season one, episode six) he likes to throw. The aforementioned common soldiers refer to Loras as the one who is "stabbing" Renly, thus seeing the latter as the effeminate part of the couple.

These ascribed gender roles become more apparent when both are together. In episode five of the first season we witness a scene in which Renly's chest and armpit are shaven by Loras. This is a key scene not included in the books. Loras shaves his lover because he prefers him hairless, clearly marking Renly as the more effeminate one of the two. This is even emphasized through their conversation, in which Renly complains that his brother treats him as if he were a spoiled child rather than a man. This opinion is not challenged by his lover Loras, who even insists that his own sword skills are the result of his continuous training, something which Renly never attempted nor ever will. Loras is clearly the driving force in this conversation and Renly appears subordinate. It is also Loras who is the first to speak to Renly about the possibility of kingship although he is only fourth in line of succession. Loras proceeds to describe Renly's qualities as king, which are exclusively non-physical in nature: "People love you. They love to serve you, because you are kind to them. They want to be near you. You are willing to do what needs to be done. But you don't gloat over it. You don't love killing. Where is it written that power is the sole province of the worst? [...] You would be a wonderful king." After that, Loras engages in fellatio, directly proving to Renly the willingness of others to serve him. In the scene, Renly is also cut by Loras. While he almost panics out of surprise, Loras forces him to look at the blood because he "need[s] to get used to it." Renly, although a good king in Loras' opinion, still has to attain at least some masculine characteristics for a possible struggle for the throne.

In this scene, we witness a break from medieval designations: Loras's masculine behavior conflicts with his assumption of the "female" role in fellatio. This sex-act can be seen as a way for the series to nuance the clear-cut masculine and feminine gender roles assigned to Loras and Renly and show a more diverse rendering of the homosexual couple on TV. However, the fellatio serves foremost as a metaphorical oath of fealty from a loyal subject to his king while at the same time providing the viewer with provocative entertainment. The act is a transgressive moment, not just because of the sexual aspect, but also because its submissive nature challenges heteronormative values of active masculine and passive, squeamish feminine. Still, the unexpected change of roles, and thereby the challenge, is only superficial. Loras is still the active part and the act was neither requested by the newly-made throne claimant nor was he asked for his approval.

Another scene in the series further emphasizes the heteronormative nature of Loras' and Renly's relationship. In season two, episode three, Loras withholds sex from Renly with the order that he sleep with his newly-wed wife, who happens to be Loras' sister, because his "vassals begin to snigger behind [his] back." Again, Loras is the active force while Renly seems intimidated by the presence of his wife. The depiction of their relationship is strongly reminiscent of almost every other contemporary homosexual couple on television, regardless of the setting, and marks a clear contrast to what has been established about medieval codes. Here, the role during the physical act is not as important in regendering one of the characters as feminine as it was, for instance, in the case of John/Eleanor Rykener; what matters is social behavior and authority.

A specifically medieval feature in this set-up is of course accessibility to the throne and subsequently the inversion of the roles of servant and master. It is only Renly, through his relation to the current king, who would be able to become king. Therefore, he is pushed into the role of leader rather than following Loras in his own attempt for the crown. The fellatio has to be seen exactly in this context. Although scheming for personal interests is one of the most prominent features of *Game of Thrones*, it seems that Loras Tyrell is genuinely attached to Renly. When Renly is killed in episode five of season two, Loras is seen standing beside his corpse with apparent and sincere affection, saying once more that he would have been "a good king." Overall, the main features of the homosexual relationship of Renly and Loras in *Game of Thrones* function along the same lines as those in other television series.

An interesting change in characterization occurs after the relationship has ended due to the premature death of Renly Baratheon. Afterwards, Loras Tyrell's feminine side is much more emphasized, while his qualities as a fighter become less important. In episode five of season three Loras starts an affair with a squire by the name of Oliver, who is clearly playing the dominant role during their encounter. He pushes Loras on the bed and also seems to have been the initiator of their affair in the first place, since Loras asks him how he knew that he wanted him. In this scene, Renly's unexpected death appears to have been completely forgotten by Loras. Rather, sex becomes the predominant factor in the series' depiction of homosexuality, a simplification that was also criticized in *Queer as Folk*. Subsequently, homosexuality is almost completely reduced to a sexual component.

The feminization of Loras' character can also be seen in a conversation between him and his soon-to-be wife Sansa Stark in episode six of the third season. The scene starts with Loras correcting Sansa that what he is wearing is not a pin, as she thought, but a brooch, thereby establishing his fashion sense. They then speak of the proposed match, to which Loras says: "I've dreamed of a large wedding since I was quite young. The guests, the food, the tournaments." Only after a long pause and after he becomes aware of Sansa's look does he add "and the bride, of course [laughs]." He then proceeds to describe his bride's wedding gown as he imagines it. This scene, as well as the love scene with Oliver, is not featured in the books. Loras' fighting skills and valor become unimportant, replaced by a sense of fashion and enjoyment of pompous feasts. One is instantly reminded of Robert's mocking remarks about Renly's fondness for balls and masquerades. Indeed it seems that Loras has taken the role of the effeminate gay male previously given to Renly.

Here, Loras' role is inverted in order to be able to present homosexuality continuously in accordance with the viewer's expectations. Again, it is not one's role during intercourse which defines a person as female or male, but one's overall behavior. In episode five of season one, Loras is introduced into the series as a participant in a joust, riding on a white mare, which initially outlines his character as soft, even feminine, especially compared to his opponent who rides a strong black war-horse. After Loras emerges victorious, however, it is revealed that his mare was in heat, therefore distracting the war-horse and giving Loras the advantage. This deconstruction of the good-looking, gallant knight who always fights with honor is also found in the novels and changes Loras' character into a cold and calculating schemer; not at all a soft, feminine knight of flowers. The scenes with Oliver and Sansa that mark Loras as an effeminate character are, interestingly enough, only present in the series. The overall purpose is to keep *Game of Thrones* as "provocative entertainment." For this, homosexuality has to stay visible within the series solely by means of modern concepts of gay masculinity. Again the series uses a neomedieval approach toward homosexuality, orienting itself toward heteronormative gender roles as we encounter them in other, non-medieval, television series with gay characters.

Game of Thrones is an essentially neomedieval product. It constructs a medieval setting by applying iconic images, but addresses contemporary concerns about homosexuality by reinscribing traditional, heteronormative gender roles. Same-sex love and sexual contact in particular serves as "provocative entertainment" for the viewer without threatening heteronormative values. The importance of these explicit scenes, all of which are not in the source material, for the overall plot of *Game of Thrones* is limited at best and, as in the novels, the show would have made sense without them. The addition of Oliver serves the purpose of continuously providing "provocative entertainment" after Renly dies. Loras' affection for Renly and any feeling of loss seem to be forgotten quickly, in favor of carnal pleasure. The inversion of Loras' masculine features emphasizes the decreasing importance of his character for the show. He is being degraded to meet the audience's expectation of a homosexual character according to modern clichés. The only other function of homosexuality lies in nuancing the personalities of straight characters such as Olenna Tyrell and Tywin Lannister. Their attitudes toward homosexuality mark them as generally open- or narrow-minded for the contemporary, liberal viewer and delineates pseudo-modern characters from "medieval" ones.

Neomedievalism disregards mere historical authenticity and establishes a medieval simulacrum in which basically anything is possible. The creators of *Game of Thrones* chose to alter their source material with regard to the depiction of homosexuality, turning it into primarily "provocative entertainment." Loras and Renly, although part of the main cast, are not at the center of *Game of Thrones* and the changes made to their characters influenced the overall plot marginally. Ultimately, one is left to wonder if truly provocative entertainment would not have been to depict two homosexual characters with a disregard for heteronormative values and gender roles, showing them both, for instance, as strong-willed leaders. However, it seems that Westeros is too close to our present time to accommodate such a concept.

Notes

1. Steven Capsuto, *Alternate Channels: The Uncensored Story of Gay and Lesbian Images on Radio and Television* (New York: Ballantine, 2000), 91ff.
2. Bonnie J. Dow, "*Ellen*, Television, and the Politics of Gay and Lesbian Visibility," *Critical Studies in Media Communication* 18 (2001): 124.
3. Guillermo Avila-Saavedra, "Nothing Queer about Queer Television: Televised Construction of Gay Masculinities," *Media, Culture & Society* 31 (2009): 11f.
4. Dow, "Ellen," 132.
5. Fred Fejes, "Making a Gay Masculinity," *Critical Studies in Media Communication* 17 (2000): 115.
6. Avila-Saavedra, "Nothing queer," 8.
7. Wendy Peters, "Pink Dollars, White Collars: *Queer as Folk*, Valuable Viewers, and the Price of Gay TV," *Critical Studies in Media Communication* 28 (2011): 196.
8. Amanda Latz and Thalia Mulvihill, "Heteronormativity in *Queer as Folk* and *The L Word*," *Academic Exchange Quarterly* 11 (2007): 157.
9. Fejes, "Gay Masculinity," 115.
10. Peters, "Pink Dollars," 197.
11. Jane Arthurs, "*Sex and the City* and Consumer Culture: Remediating Postfeminist Drama," *Feminist Media Studies* 3 (2003): 84.

12. Peters, "Pink Dollars," 199.

13. Dana Frei, "Challenging Heterosexism from the Other Point of View: Representations of Homosexuality in Present-Day Television Series," Schweizerisches Archiv für Volkskunde 103 (2007): 101.

14. "Where We Are on TV," GLAAD, accessed July 28, 2014, http://www.glaad.org/publications/whereweareontv12.

15. "George R. R. Martin Interview GAME OF THRONES," Christina Radish, accessed July 28, 2014 http://collider.com/george-r-r-martin-interview-game-of-thrones/#x4jIehhoFA770kIo.99.

16. Massively Multiplayer Online Role-Playing Game.

17. "GRRM Interview Part 2: Fantasy and History," James Poniewozik, accessed July 28, 2014, http://entertainment.time.com/2011/04/18/grrm-interview-part-2-fantasy-and-history/.

18. "George R.R. Martin: The Complete Unedited Interview," Charlie Jane Anders, accessed July 28, 2014, http://observationdeck.io9.com/george-r-r-martin-the-complete-unedited-interview-886117845.

In another interview George R.R. Martin referred to the Middle Ages as his "model" for the books. "George R. R. Martin Interview, Part 1: Game of Thrones, from Book to TV," James Poniewozik, accessed July 28, 2014, http://entertainment.time.com/2011/04/15/george-r-r-martin-on-game-of-thrones-from-book-to-tv/.

19. For an overview see: Pamela Clements and Carol L. Robinson, "Neomedievalism Unplugged" *Studies in Medievalism: Corporate Medievalism* 21 (2012): 191–205. Bernd and Kevin Moberly argue that the focus on audience-expectations results in a cliché depiction of the Middle Ages which is "a version of the medieval which is more medieval than the medieval." See Brent Moberly and Kevin Moberly, "Neomedievalism, Hyperrealism and Simulation" *Studies in Medievalism: Defining Medievalism(s)* 19 (2010): 15.

20. I am indebted to Karolyn Kinane for pointing out the connection between the mythical elements in neomedieval works and what can be termed "the literary Middle Ages." I am also indebted to her for sending me a draft of her paper on "New Age and Neopagan Medievalisms" forthcoming in *Handbook of Medieval Afterlives in Contemporary Culture* (Bloomsbury, 2015), which, among other focuses, discusses the authenticity of the literary Middle Ages for these religious movements. See also Karolyn Kinane, "Intuiting the Past: New Age and Neopagan Medievalisms," *Relegere: Studies in Religion and Reception* 3, no. 2 (2013): 225–248, esp. 239ff.

21. *The Anglo-Saxon Chronicle*, trans. and ed. Michael Swanton (New York: Routledge 1998), p. 55.

22. "Game of Thrones is more brutally realistic than most historical novels," Tom Holland, accessed July 28, 2014, http://www.theguardian.com/tv-and-radio/2013/mar/24/game-of-thrones-realistic-history.

23. Sir Thomas Malory, *Le Morte Darthur: A Norton Critical Edition*, ed. Stephen H. A. Shepherd (New York: W.W. Norton, 2004), p. 158 speaks, for instance, of Lancelot striking down five knights with one thrust of his spear.

24. Michel Foucault places the emergence of the notion of "sexuality" in the period following the eighteenth century. Michel Foucault, *The History of Sexuality Volume 1: An Introduction*, trans. Robert Hurley (Harmondsworth: Penguin, 1981), 38.

25. See for an introduction to the topic within medieval research Allen J. Frantzen, *Before the Closet: Same-Sex Love from Beowulf to Angels in America* (Chicago: University of Chicago Press, 1998), 1ff.

26. John Boswell, *Christianity, Social Tolerance, and Homosexuality: Gay People in Western Europe from the Beginning of the Christian era to the Fourteenth Century* (Chicago: University of Chicago Press, 1980), 180.

27. Connor McCarthy, ed., *Love, Sex and Marriage in the Middle Ages: A Sourcebook* (London: Routledge, 2004), 65f.

28. *Ibid.*, 45.

29. *Ibid.*, 127.

30. See for example Frantzen, *Before the Closet*.

31. W. Mark Ormrod, "The Sexualities of Edward II" in *The Reign of Edward II: New Perspectives*, eds. Gwilym Dodd and Anthony Musson (York: York Medieval Press, 2006), 28ff.

32. Rogerus de Hoveden, *Gesta Henrici Secundi: Gesta regis Henrici Secundi Benedicti abbatis.*

The Chronicle of the Reigns of Henry II, and Richard I. A.D. 1169–1192. Known Commonly under the Name of Benedict of Peterborough, ed. William Stubbs, vol. 49 of *Rolls Series* (London: Longmans, Green, Reader, and Dyer, 1867), vol. 2, 7.

Homosexual tendencies were only later ascribed to the relationship between Richard and Philipp, for example in the play "The Lion in Winter" by James Goldman (1966).

33. Boswell, *Homosexuality*, 334.

34. David W. Marshall, "Neomedievalism, Identification, and the Haze of Medievalisms," *Studies in Medievalism: Defining Medievalism(s) II* 20 (2011): 24.

35. One other instance of a homosexual act has to be categorized differently. In season three, episode three Theon Greyjoy is almost raped during an attempted escape by one of his former guards. However, neither lust nor affection is the motivator for the guard in this instance. Rather, the act is intended to establish Theon's subordination and proves the almost limitless power the guard has over him in this moment.

36. George R. R. Martin, *A Storm of Swords* (New York: Bantam Books, Mass Market Edition, 2011), 848.

Bibliography

Anders, Charlie Jane. "George R.R. Martin: The Complete Unedited Interview." Accessed July 28, 2014. http://observationdeck.io9.com/george-r-r-martin-the-complete-unedited-interview-886117845.

The Anglo-Saxon Chronicle, trans. & ed. Michael Swanton. New York: Routledge 1998, p. 55.

Arthurs, Jane. "Sex and the City and Consumer Culture: Remediating Postfeminist Drama." *Feminist Media Studies* 3 (2003): 83–98.

Avila-Saavedra, Guillermo. "Nothing Queer about Queer Television: Televised Construction of Gay Masculinities." *Media, Culture & Society* 31 (2009): 5–21.

Boswell, John. *Christianity, Social Tolerance, and Homosexuality. Gay People in Western Europe from the Beginning of the Christian Era to the Fourteenth Century*. Chicago: University of Chicago Press, 1980.

Capsuto, Steven. *Alternate Channels: The Uncensored Story of Gay and Lesbian Images on Radio and Television*. New York: Ballantine, 2000.

Clements, Pamela, and Carol L. Robinson. "Neomedievalism Unplugged" *Studies in Medievalism: Corporate Medievalism* 21 (2012): 191–205.

Dow, Bonnie. "Ellen, Television, and the Politics of Gay and Lesbian Visibility." *Critical Studies in Media Communication* 18 (2001): 123–140.

Fejes, Fred. "Making a Gay Masculinity." *Critical Studies in Media Communication* 17 (2000): 113–116.

Foucault, Michel. *The History of Sexuality Volume 1: An Introduction*. Trans. Robert Hurley. Harmondsworth: Penguin, 1981.

Frantzen, Allen J. *Before the Closet: Same-Sex Love from Beowulf to Angels in America*. Chicago: University of Chicago Press, 1998.

Frei, Dana. "Challenging Heterosexism from the Other Point of View: Representations of Homosexuality in Present-Day Television Series." *Schweizerisches Archiv für Volkskunde* 103 (2007): 83–103.

GLAAD. "Where We Are on TV." Accessed July 28, 2014. http://www.glaad.org/publications/whereweareontv12.

Holland, Tom. "Game of Thrones is more Brutally Realistic than most Historical Novels." Accessed July 28, 2014. http://www.theguardian.com/tv-and-radio/2013/mar/24/game-of-thrones-realistic-history.

Kinane, Karolyn. "Intuiting the Past: New Age and Neopagan Medievalisms." *Relegere: Studies in Religion and Reception* 3, no.2 (2013): 225–248.

Kinane, Karolyn. "New Age and Neopagan Medievalisms" (forthcoming).

Latz, Amanda, and Thalia Mulvihill. "Heteronormativity in Queer as Folk and The L Word." *Academic Exchange Quarterly* 11 (2007): 156–160.

Malory, Sir Thomas. *Le Morte Darthur. A Norton Critical Edition*, ed. Stephen H. A. Shepherd. New York: W.W. Norton, 2004.

Marshall, David W. "Neomedievalism, Identification, and the Haze of Medievalisms." *Studies in Medievalism: Defining Medievalism(s) II* 20 (2011): 21–34.

Martin, George R. R. *A Storm of Swords*. New York: Bantam Books, Mass Market Edition, 2011.

Mayer, Lauren S. "Unsettled Accounts: Corporate Culture and George R. R. Martin's Fetish Medievalism." *Studies in Medievalism: Corporate Medievalism* 21 (2012): 57–64.

McCarthy, Connor, ed. *Love, Sex and Marriage in the Middle Ages: A Sourcebook*. London and New York: Routledge, 2004.

Moberly, Bernd and Kevin Moberly. "Neomedievalism, Hyperrealism and Simulation." *Studies in Medievalism: Defining Medievalism(s)* 19 (2010): 12–24.

Ormrod, W. Mark. "The Sexualities of Edward II," In *The Reign of Edward II: New Perspectives*, ed. Gwilym Dodd and Anthony Musson, 22–47. York: York Medieval Press, 2006.

Peters, Wendy. "Pink Dollars, White Collars: Queer as Folk, Valuable Viewers, and the Price of Gay TV." *Critical Studies in Media Communication* 28 (2011): 193–212.

Poniewozik, James. "George R. R. Martin Interview, Part 1: Game of Thrones, from Book to TV." Accessed July 28, 2014. http://entertainment.time.com/2011/04/15/george-r-r-martin-on-game-of-thrones-from-book-to-tv/.

Poniewozik, James. "GRRM Interview Part 2: Fantasy and History." Accessed July 28, 2014. http://entertainment.time.com/2011/04/18/grrm-interview-part-2-fantasy-and-history/.

Radish, Christina. "George R. R. Martin Interview GAME OF THRONES." Accessed July 28, 2014. http://collider.com/george-r-r-martin-interview-game-of-thrones/#x4jIehhoFA770kIo.99.

Rogerus de Hoveden. *Gesta Henrici Secundi: Gesta regis Henrici Secundi Benedicti abbatis. The Chronicle of the Reigns of Henry II, and Richard I. A.D. 1169–1192. Known Commonly under the Name of Benedict of Peterborough*, edited by William Stubbs. Vol. 49 of Rolls Series. London: Longmans, Green, Reader, and Dyer, 1867.

About the Contributors

Stephanie L. **Coker** is an assistant professor of French at Oral Roberts University in Tulsa, Oklahoma, where she teaches courses on French language, literature, and culture. Her research focuses on how French authors have retold Jeanne d'Arc's story from the Hundred Years War to the period following World War II.

Melissa Ridley **Elmes** is a doctoral candidate in English at the University of North Carolina at Greensboro. Her primary research focus is the study of feasts and feasting in medieval British literature. Recent publications have focused on cultural transmission of the Arthurian legend, Chaucer's *Parliament of Fowls*, and the figure of Melusine.

Sandy **Feinstein** is an associate professor of English at Penn State Berks. Her scholarship covers a wide range: from the literature of the Middle Ages and early modern period in England and Europe to pedagogy and poetry. Recent publications include "Longevity and the Loathly Ladies in Three Medieval Romances " and, with Neal Woodman, "Shrews, Rats, and a Polecat in the Pardoner's Tale."

Tara **Foster** is an associate professor of French at Northern Michigan University. Her research interests range from Old French narrative and hagiography to medievalism and popular culture. Her publications include articles on Clemence of Barking, Juana of Castile, and literary representations of the Third Crusade.

Torben R. **Gebhardt** teaches at the Westfälische Wilhelms-University Münster. His doctoral dissertation focuses on the group affiliation of monarchs in England and Germany in the period between 1016 and 1138.

Michael W. **George** is an associate professor of English at Millikin University, where he teaches English literature through the 18th century, publishing, and fantasy. His research interests include humor in medieval literature and representations of the environment in early English literature.

Karolyn **Kinane** is an associate professor of English at Plymouth State University. She publishes and teaches on mysticism, contemplative pedagogy, and medievalism. She edited a special issue of *Relegere* on New Age and Neopagan medievalisms and, with Michael Ryan, *End of Days: Essays on the Apocalypse from Antiquity to Modernity* (McFarland, 2009).

Shannon **McSheffrey** is a professor of history at Concordia University. Her research focuses on how late medieval Londoners used law, legal records, and legal archives. She has published four books and numerous articles on gender roles, law, civic culture, marriage, literacy, heresy, and popular religion in late medieval and Tudor England.

Elysse T. **Meredith** received a Ph.D. in medieval studies from the University of Edinburgh. Associated with English literature, French language and literature, and art history, her doctoral thesis examines the use of color and costume in 14th- and 15th-century Arthuriana.

Meriem **Pagès** is an associate professor of English at Keene State College. While her primary research interest lies in the representation(s) of Islam in medieval Europe, she has also published on contemporary popular medievalism and pedagogy.

Evan **Torner** is an assistant professor of German studies at the University of Cincinnati. He has written on East Germany, critical race theory, DEFA Indianerfilme, science fiction, transnational genre cinema, and game studies. Recent work includes the *Handbook of East German Cinema: The DEFA Legacy,* co-edited with Henning Wrage, and *Solidarity? Race in East German Cinema.*

Angela Jane **Weisl** is a professor of English at Seton Hall University. She is the author of *The Persistence of Medievalism: Narrative Adventures in Contemporary Culture* and *Conquering the Reign of Femenye: Gender and Genre in Chaucer's Romance.* With Tison Pugh, she wrote *Medievalisms: Making the Past in the Present* and co-edited the *MLA Approaches to Teaching Chaucer's Troilus and Criseyde and the Shorter Poems.*

Index

www.ingramcontent.com/pod-product-compliance
Ingram Content Group UK Ltd.
Pitfield, Milton Keynes, MK11 3LW, UK
UKHW020615180726
13836UKWH00010B/2490